The Pomegranate

The Pomegranate

M.K. SHEIKH
M.Sc. (Agri.), Ph.D. (Hort.)
Professor & Head
Department of Horticulture
College of Agriculture
University of Agricultural Sciences
Bijapur (Karnataka) INDIA

CBS
CBS Publishers & Distributors Pvt. Ltd.
New Delhi • Bengaluru • Chennai • Kochi • Mumbai • Pune
Hyderabad • Kolkata • Nagpur • Patna • Vijayawada

The Pomegranate

ISBN: 978-81-239-2687-2

First CBS Reprint: 2015

Published by:
Satish Kumar Jain for CBS Publishers & Distributors Pvt. Ltd.,
4819/XI Prahlad Street, 24 Ansari Road, Daryaganj, New Delhi - 110002
delhi@cbspd.com, cbspubs@airtelmail.in • www.cbspd.com
Ph.: 23289259, 23266861, 23266867 • Fax: 011-23243014

Corporate Office: 204 FIE, Industrial Area, Patparganj, Delhi - 110 092
Ph: 49344934 • Fax: 011-49344935
E-mail: publishing@cbspd.com • publicity@cbspd.com

Branches:

- ***Bengaluru:*** 2975, 17th Cross, K.R. Road, Bansankari 2nd Stage, Bengaluru - 70
  Ph: +91-80-26771678/79 • Fax: +91-80-26771680
  E-mail: cbsbng@gmail.com; bangalore@cbspd.com
- ***Chennai:*** No. 7, Subbaraya Street, Shenoy Nagar, Chennai - 600030
  Ph: +91-44-26681266, 26680620 • Fax: +91-44-42032115
  E-mail: chennai@cbspd.com
- ***Kochi:*** 36/14, Kalluvilakam, Lissie Hospital Road, Kochi - 682018
  Ph: +91-484-4059061-65 • Fax: +91-484-4059065
  E-mail: cochin@cbspd.com
- ***Mumbai:*** 83-C, Dr. E. Moses Road, Worli, Mumbai - 400018
  Ph: +91-9833017933, 022-24902340/41 • E-mail: mumbai@cbspd.com
- ***Pune:*** Bhuruk Prestige, Sr. No. 52/12/2+1+3/2,
  Narhe, Haveli (Near Katraj-Dehu Road Bypass), Pune - 411041
  Ph: +91-20-64704058/59, 32342277 • E-mail: pune@cbspd.com

Representatives:

- Hyderabad: 0-9885175004
- Kolkata: 0-9831437309, 0-9051152362
- Nagpur: 0-9021734563
- Patna: 0-9334159340
- Vijayawada: 0-9000660880

Printed at:
India Binding House, Noida (UP)

Preface

Pomegranate cultivation is one of the most remunerative farming enterprises in India. It is grown in Maharashtra, Karnataka, Andhra Pradesh, Gujarat on a large scale. The total pomegranate production in the world is 10 lakh tonnes. India produces 5 lakh tonnes but export only 5000 tonnes, whereas Spain produces 1 lakh tonne and export 75,000 tonnes and, their export ends by December. India has the scope to export from January to June to European countries. Farmers of pomegranate are advised to take Hasta bahar flowering by pruning in August and harvesting in March, to prevent Bacterial blight (*Xanthomonas axonopodis* cv. *Punicae*) problem.

This book covers all aspects viz., soil, water, climate, propagation, varieties, the establishment of a garden, pruning, pest and disease, nutrition, value-added products, wine-making etc.

I sincerely hope that this book will prove helpful to pomegranate growers and extension personnel in Maharashtra, Karnataka and Andhra Pradesh.

Dr. M.K. Sheikh

Contents

Chapter 1

Introduction

POMEGRANATE (PUNICA GRANATUM L.)

The Pomegranate (Punica granatum L.) is one of the ancient and highly praised favourite fruits. It is commercially grown, apart from India, in a number of countries for its sweet-acidic fruits, which provide a cool refreshing juice, and is valued from its medicinal properties. Its popularity is also due to the ornamental nature of the plant which bears bright red, very attractive flowers. To highlight its importance, it was chosen as a symbol of the 18th International Horticultural Congress held during 1970, showing it in a basket.

Pomegranate can successfully be cultivated even under conditions of severe drought and frost. The fruit rind, juice, leaf and roots are used in the preparation of various Ayurevedic medicines. The Juice and seeds also contain large quantities of tanin and agolic acid, which is essential in curing several diseases.

NUTRITIVE VALUE

(Per 100 gm of edible portion)

Moisture	-	78.00 gms.
Protein	-	1.60gms.
Fat	-	0.10 gms.
Minerals	-	5.10 gms.
Phosphorous	-	0.07 gms.
Iron	-	0.30 mg.
Riboflavin	-	100.00 mg.
Vitamin 'C'	-	16.00 mg.

In the recent past pomegranate cultivation gained momentum due to the production technologies developed by the scientists. Among the different states Maharashtra is leading in pomegranate production. It is also grown commercially In Karnataka, Andhra Pradesh, Rajasthan, Gujarat and Tamil Nadu. As there is tremendous scope for exports, the central and state governments are providing several subsidies. Subsidies are being provided in several states for the installation of drip irrigation in pomegranate orchards. For competing in International trade our cultivators have to produce pomegranate fruits of International quality standards. To achieve these standards the cultivation details are furnished hereunder.

Pomegranate (Punica granatum L.) belongs to the natural order *Punicaceae. Punica* perhaps is the only known genus of this family, which includes large shrubs over small trees with 2 species. *Punica Protopunica* Balf. S is found wild in Socotra Island and the other *Punica granatum* is cultivated in tropical and sub-tropical parts of the world. Punica granatum has been classified into two sub-species *Chorocarpa* and *Porphyrocarpa,* each having 2 varieties. These sub-species have been established on the basis of the colour of the ovary, a stable feature which is retained when they are reproduced by seeds. Sub-species *Chlorocarpa* is mainly found in the *Trancaucasus,* whereas, the second sub-species *Porphyrocarpa* is mainly distributed in central Asia.

Pomegranate (*Punica granatum*) belonging to the family punicacea is one of the favourite table fruits of tropical and sub-tropical regions. The fruit is grown in Iran and is extensively cultivated in Mediterranean countries like Spain, Morocco, Egypt, Iran, Afghanistan and Baluchistan. It is also grown to some extent in Burma, China, Japan, USA (California), USSR and India. It is also found growing in Bulgaria and in Southern Italy.

Pomegranate (*Punica granatum*) is native to Persia and

surrounding area. Pomegranate is associated with the most ancient civilization in the Middle East and later it naturalized over the whole Mediterranean region.

It flourished particularly well in Spain and the city of Granada where high quality pomegranate are grown.

Pomegranate is quite suitable to hot and arid regions and therefore it also spread to India and china. Spanish missionaries brought the pomegranate to the new world including Mexico and Californaia in U.S.A.

India is the World's leading producing country of pomegranates. Approximately 54,735 hectares area is cultivated for pomegranate. The details of area and production in different states is given below.

Table-1 : Pomegranate production in India

Sl. No.	State	Area under Cultivation (ha)	Production in tons	Productivity Tons/ha
1	Maharashtra	40,970	4,09,700	10.0
2	Karnataka	8,454	49,059	5.80
3	Gujarat	2,988	21,987	7.40
4	Andhra Pradesh	2,000	16,000	8.00
5	Rajasthan	363	4,102	11.30
	Total	54,755	5,00,848	8.60

Its cultivated area is as much as 71,737 ha which are under cultivation in Mahrashtra state. However, only 40,970 ha are in the productive stage. At present, Maharashtra is the, main pomegranate producing state. Area under pomegranate in this state has increased from 31,549 ha in 1996-97 to 40,970 ha in 1999-2000.

Next to Maharashtra, maximum pomegranate cultivation

takes place in Karnataka where the area has increased from 7,584 ha in 1996-97 to 8,454 ha in 1999-2000.

Pomegranate is also cultivated in other parts of the world. However, India remains the leading producing country (5,00,848 tons). Other countries well-known for pomegranate production are Spain (1,00,000 tons approx.) and Iran (1,00,000 tons approx).

Pomegranate is also quite a nutritive fruit as it is rich in riboflavin, niacin, and ascorbic acid and also in potassium.

In India, it is largely used as desert fruit and its fresh juice is also very much relished. In foreign countries, it is commonly used for decorative purposes its fresh kernels are used for garnishing desserts and salads. Pomegranate juice is mainly used for making Grenadine (a type of syrup). Which is further used for flavouring of mixed drinks, ice creams or desserts.

Edible quality of Indian Pomegranates especially of soft seeded ones, i.e., Ganesh, G-137, Mridula, Jalore seedless, Ruby, etc, seems to be much better, as their T.S.S. varies from 15 to 17 Degree brics and acids between 0.3% to 0.45% whereas Spanish Pomegranate have T.S.S. between 17 to 18 Degree Brics and acids from 1.5% to 1.8%. Probably this is the reason why Spanish Pomegranate are more used for decoration, garnishing or for syrup making, whereas Indian Pomegranates are more used for fresh fruit juice purposes.

Concentrated pockets of Pomegranate in India

Table 2:Concentrated pockets of pomegranate growing areas

State	District	Productive area in ha	Production in tons
Maharashtra	Solapur	15,157	1,51,570
	Nasik	10,800	1,08,000
	Sangli	5,300	53,000
	Ahmednagar	2,615	26,150
Karnataka	Bijapur	8,454	49,059
Andhra Pradesh	Anantpur	2,000	16,000

[Source : Departments of Horticulture, Maharashtra, Pune; A.P, Hyderabad & NHB, Gurgaon]

Maharashtra state accounts for 78% of the total area in India and 84% of the total production in the country. Pomegranate is concentrated in 4-5 districts and prominent districts along with their area under pomegranate are as follows:

The prominent pockets where pomegranate area are concentrated are Solapur, Nasik, Sangli, and Ahmednagar in Maharashtra and Bijapur in Karnataka state. In Andhra Pradesh also concentrated area for Pomegranate cultivation has emerged in Anantput district.

Maximum area of Pomegranate is occupied by Ganesh variety (approx. 90%) and other important varieties include G-137, Mirdula, Bhagwa and Arkata. All these cultivars are soft seeded having high T.S.S and low acids with light pink coloured arils to dark red coloured arils.

Export scenario of fresh Pomegranate from India

Detailed data on exports of fresh pomegranates from India are presented in Table 3. This shows India plays a significant role in the global market. Recently exports from India of Pomegranates has extended to several countries. The figures indicate that during 1999-2000 as much as 5,726 tons of pomegranates were exported for a total of Rs. 11.5 croers. However, during 2000-2002, exports declined to 4,455 tons for a total of Rs. 9.91 crores.

Most notable point is that the exports for the last four years are almost varying between 4,000 tons to 5,500 tons. These exports form only approx. 1.0% of the total production whereas Spain produces approx. 1,00,000 tons of pomegranate and exports 75% of the production.

A critical analysis of export figures show that during 2000-2001, maximum exports of pomegranate were Gulf countries and particularly UAE (2,922 tons), Saudi Arabia (170 tons), Bahrain (193 tons) and Kuwait (65 tons).

During 1999-2000 exports of pomegranate to Middle East countries were much higher i.e., 3,159 tons to UAE; 1,232 tons to Saudi Arabia, 160 tons to Bahrain and 139 tons to Kuwait. This suggests that exports to this region can be considerably accelerated as there seems to be excellent potential.

As far as exports of fresh pomegranate to SAARC countries are concerned, sizeable qauality is exported to Bangladesh, Sri Lanka and Pakistan (Table 3). In these countries also there is a good scope for stepping up exports of fresh pomegranate.

After Gulf and SAARC countries, next in terms of quantum of exports are the European countries. U.K. is the principal importer buying 161 tons in 2000-2001. However, in earlier years, U.K. imported in much larger quantities, i.e.., 358 tons in 1997-98; 273 tons in 1998-99 and 241 tons in 1998-99. Similarly, countries like France, Germany and Switzerland imported 35 tons, 48 tons and 21 tons respectively during 1998-99. More recently in 2000-2001, our exports to Europe have diversified as we exported 140 tons to Italy, 80 tons to Netherlands and 31 tons to Belgium. This shows there is wide acceptability for Indian pomegranate and this needs to be boosted by nuturing the markets.

There is renewed interest in exploration to Far East countries where there is a good potential for markteing. Pomegranate was exported to the Phillipines (35 tons in 1998-99), Singapore (22 tons in 1999-2000) and Hong Kong (15 tons in 1998-99).

There are very little scope for exports to African countries except in South Africa, where 47 tons of fresh pomegranates were exported during 2000-2001. This needs to be further viewed in terms of marketing it in future. However, there has been negligible exports to other countries in this region.

Regarding North America, there are scope for exports to Canada. During 1999-2000, 60 tons of fresh pomegranates were exported and during 2000-2001, tons, 51 tons were exported. Price realization is also quite high and considering longer shelf life of pomegranates, countries in this continent like Brazil, Mexico, Trinidad, Barbados and Bahamas where small quantities of fresh pomegranates have been exported, efforts are needed to expand the market there.

World market for pomegranates

Production

Data on production of pomegranate is not available in F.A.O. statistics. However, old data on website, shows that the present production of pomegranate is of the order of 10,00,000 tons.

Out of one million tons production, India is the leading producer of pomegranate. With a production of 5,00,848 tons, Spain and Iran produce approx, 1,00,000 tons each. Rest of the produce comes form Egypt, Israel, Jordan, Lebanon, Tunisia, Pakistan, Thailand, Chile, Peru and U.S.A.

World Trade

As per estimates, total world trade is around 1,00,000 tons to 1,20,000 tons per year. However, no official statistics are available from F.A.O. or any other source. Out of this quantity, Spain alone exports about 70,000 to 75,000 tons per year. Spain is a dominant player in world exports. Next to Spain, Iran exports approx 15,000 tons, whereas India exports olnly 5,000 tons per year. Pomegranate from India and Iran are almost available throughtout the year while Spanish pomegranates are in the peak in the months of September to December. Spain continues to dominate EU market due to advantage in freight cost over Iran and India. Iran also exports to Europe during November to March, However in Middle East market both Iran and India are the main exporters.

Important aspects of the pomegranate trade is that Spain produces only about 10% of the worlds production but has more than 50% share in world exports. Whereas Indian produces almost 50% of the world pomegranates but has a mere 5% share in world exports.

a. *Importing countries*

The main importing countries are in Europe and Middle East. Imports of the order of 80,000 to 90,000 tons/year are done by EU market, comprising of U.K. France, Germany, The Netherlands, Sweden, Switzerland, Denmark, Belgium and Italy. Whereas Gulf countries import to the tune of 15,000 tons to 25000 tons/years. The major Middle East importing countires are Saudi Arabia, U.A.E. Bahrain, Kuwait, Qatar, Oman, etc.

b. *Quality requirements of importing countries*

The desireable fruit characters for export purposes are detailed below:

i Dark rose pink colour of fruit

ii. Fruit weight of around 200-400 gm.

iii. Round shape of fruit.

iv. Uniform size and shape of the fruit in pack or box.

v. Dark rose pink arils

vi. Softness of the seeds

vii. Higher sugar contents of about 16 °B to 17 °B.

viii. Freedom from scars, russeting, disease spots, insect injury, and scratches.

ix. Smooth cutting at the stem end.

x. Bracts/calyx without any damage and having freshness.

xi. Pleasant flavour and aroma

The wild pomegranate fruit is filled with angular hard seeds covered with a juicy, pink or yellowish white, sweet astringent acid pulp. The fruit is used by sun-drying the seeds along with the pulp. Which constitute the product 'anardana' a valuable commodity used as a condiment and for other culinary purposes. Methods have been standardized for the commercial exploitation of anardana. The seeds have oestrogenic activity due to the presence of 1.00 mg oestrogene and 0.036 mg coumestrol (a non-steroidal oestrogen), 100 g-1 of seeds *etc.* The products of pomegranate such as bottled juice, syrup and jelly are highly appreciated because of their nutritive and therapeutic qualities. The surface colour of fruit varies in few commercial cultivars from yellow with a crimson cheek, from brownish red to bright red. The edible portion is the bright-red pulp surrounding the individual seeds, six anthocyanin pigments are responsible for the red color of pomegranate and identified as delphindin-3-glucoside and 3,5-diglucosde, cyaniding-3-glucoside, pelargonidin-3-glucoside and 3,5-diglucoside. Generally, there was an increase in juice pigmentation with fruit ripening. The concentration of pigments in juice obtained from matured fruits range from 50 to 100 mg anthocyanin and g-1 fresh weight of aril. The total amount of pigment in the juice was generally less in fruits with reddish skin than in those with yellow skin.

CLIMATE

Pomegranate has a versatile adaptability to wide ranging climatic condition. It can grow in the plains as well as on the hills upto an elevation of about 1840 metres. This fruit tree grows in low termperature, the tree is deciduous but in tropical and sub-tropical conditions it is evergreen or partially deciduous. The tree requires hot and dry climate during the period of fruit development and ripening. The optimum temperature for fruit development is about 38 degree C. The tree cannot produce sweet fruits unless the temperature is high for a sufficiently long period. The quality of fruit is adversely affected in humid climate.

Pomegranate is winter-hardy and drought-tolerant plant and can thrive under desert conditions. Although, it is highly drought-resistant, the plant bears well only under irrigation. Aridity and frequent anomalies of the climate cause leaf shedding and fruit cracking.

SOILS

The pomegranate can be grown on diverse soil types. The tree can be grown in soils which are considered unsuitable for most other fruit crops. It can tolerate soils which are limey and slightly alkaline. The deep loamy or alluvial soils are ideal for its cultivation though it can be grown in medium or light black soils with minimum of 60 cm depth.

Cultivars grown in India

a. Varieties

i. Ganesh

Ganesh is a selection from open pollinated seedlings of Alandi. Earlier it was known as GBG-1. This variety has pinkish yellow to reddish yellow rind colour, having soft seeds and light pink arils fruits weighing between 225 to 250 gm, with T.S.S. 16 degree B with acidity of 0.3%. This cultivar has the highest cultivated area and very popular among the farmers and is also mainly exported. The arils of this cultivar attain pink color in winter months and are whitish in warmer months.

ii. Muscut

This variety used to be cultivated in Mahrashtra and has fruit weight range of 300-350 gms. The rind is yellowish pink, the grains are medium hard and whitish pink in colour. The T.S.S. is 16 degree B with acidity of 0.5%.

At present this variety is not commercially cultivated and it not used for export.

iii. Bassein seedless

This variety is cultivated to some extent in Karnataka state. The fruit weight varies from 260 to 300 gm. The outer rind colour is red, the seeds are soft having light pink coloured arils. The T.S.S. is 16 degree B with acidity of 0.37%. There is limited area under this cultivar. Even in Bijapur district in Karnataka state, Ganesh is cultivated commercial.

iv. Jodhpur Red

This cultivar is from Rajasthan state. The fruits of this variety weigh between 180 to 220 gms. The rind colour is yellowish red with whitish pink arils and hard seeds. The T.S.S. is 15 degree B with 0.6% acidity. This variety has no commercial importance.

v. Jalore seedless

This variety is cultivated in Rajasthan and has fruits weighing between 280 to 320 gm. The rind colour is reddish yellow to pinkish yellow. The seeds are very soft and pink coloured. The T.S.S. is 15 degree B with acidity of 0.42%. The area under this variety is increasing in Rajasthan.

vi. Dholka

Dholka variety is cultivated in Gujarat. It has red coloured fruits. The seeds are soft and arils are whitish pink. T.S.S. is 14 degree B with acidity of 0.41%. The fruits weigh between 280 to 310 gm. This is grown in a limited area in Gujarat state.

b. Recent selections

i. GKVK-1

This variety is selected from Bassein seedless by UAS, Bangalore. The fruit is yellowish red in colour with soft seeds and deep red arils. The fruit weighs between 220-240 gms and has T.S.S. of 15 degree B with acidity of 0.56%.

ii. IIHR selection

This variety is also a selection from Bassein seedless made by IIHR, Bangalore. The fruit of this variety is pinkish yellow and weighs about 180 to 120 gms. It has light pink arils and soft seeds with T.S.S. of 13 degree B and acidity of 0.55%.

iii. G-137

This is a selection from open pollinated seedlings of Ganesh variety. This selection was made by MPKV, Rahuri. The fruit of this variety weighs between 250 to 280 gm, is reddish yellow in colour. The seeds are soft and arils are light pink in color. The T.S.S. of this variety is 15 degree B with acidity of 0.36%. This is the only variety which has some acreage under cultivation in Mahrashtra state.

Iv. P-23

This is also a selection fro Muscut variety. The fruit of this variety is greenish yellow with pink tinge and weighs around 370 to 400 gms. The seeds are medium hard with light pink arils. The fruit has T.S.S. of 16 degree B and acidity of 0.42%.

v. P-26

This is also a selection from Muscat variety. The rind of the fruit is greenish yellow with pink blush. The fruit weighs between 360 to 400 gms. The seeds are soft with light pink coloured arils. The T.S.S. is 15 degree with acidity of 0.41%.

vI. CO-I and Yercaud selection from TamIl Nadu state

CO-I is a selection made by TNAU, Coimbatore, having softs seeds. Similarly, Horticulture Research Station, Yercaud, selected a clone namely, Yercaud-I having medium sized fruits, easily peelable rind with soft seeds and attractive deep purple arils.

c. Hybrids

i. Mirdula

It is a selection from a cross between Ganesh and Gulsah Rose Pink varieties done by M.P.K.V., Rahuri. The fruit of this variety is red in colour and weighs between 230-270 gms with blood red arils and soft seeds, T.S.S. of 17-18 degree B and acidity of 0.47%. This variety is in great demand from farmers.

ii. Ruby

Ruby is a variety released recently by IIHR, Bangalore. This variety is a complex hybrid between Ganesh, Kabul, Yercaud and Gulsha Rose Pink varieties. The fruit weighs between 225 to 275 gms. The rind colour is red, seeds are soft with dark red coloured arils having T.S.S. of 17 degree B and acidity of 0.64^

iii. Arakta

This variety is also a selection from open pollinated population of a cross between Ganesh and a Russian variety. The fruit of this variety is dark red in colour, soft seeds, dark red colored arils and high T.S.S. It is reported that it has acreage next to Ganesh and this is the most sought-after variety among farmers.

iv. Bhagwa

This variety has attractive glossy red thick skin with blood red and blood arils. It has also soft seeds and high T.S.S. This is also very popular among the farmers and is cultivated in large areas. It has also good shelf life and tolerant to fruit cracking. The Bhagawa variety of pomegranate presently under cultivation known by different names viz. Shendria, Ashtagandha, Kesar. The variety mature in 180-190 days (1) It fetches better prices which is 2-3 times higher than Ganesh, (2) Heavy demand for export (3) Suitable for long distance market (4) Better keeping quality (15-20 days at room temperature) (5) Moderately susceptible to black spots (6) Free from blackening of arols (7) No incidence of fruit cracking (8) No fruit drop (9) It gives high yield of 30-40 kg/tree (10) Tolerant to thrips, mites.

Suitability of Indian Cultivars for exports as fresh fruit

At present almost entire export is of cultivar Ganesh grown in maximum acreage (90 %). The other varieties have small acreage namely, Arakta (5 % of area), Mridula (2-3 % area), Bhagwa (1 %) and G-137 (0.5 % area). The acceptability of varieties Arakta, Mridula, Bhagwa, etc. is much in foreign markets because of very attractive fruit and aril colour.

Foreign cultivars

a. Varieties from spain, U.S.A. and Israel

i Wonderful

This variety originated in Florida, it has deep purple red fruit with medium thick and tough rind and arils are deep crimson in colour, juicy, with medium soft seeds. Wonderful cultivar, has a T.S.S. of 17 degree B or more and titrable acidity of 1.8%. This is a leading commercial variety. The fruit matures in late September and October. This variety ships well.

ii. Granada

This is a patented early maturing variety. It originated in California as a bud mutation of Wonderful. Fruit is darker red in colour with less acidity. The fruit ripens one month earlier than wonderful but smaller in size than Wonderful. Because of early maturity it commands a premium place in the market.

iii. Ruby Red

This variety has a limited commercial importance. The fruits are of the same size as Wonderful. It has crimson purple coloured fruits. It matures at the same time and reduces splitting of fruits. This variety does not store well as Wonderful. (Courtesy APEDA, New Delhi - 2002)

PROPAGATION

For the production of quality pomegranate plants vegetative propagation is the best method. Cutting obtained from one-year-old shoots are universally used for pomegranate propagation. 9-10" long cuttings are planted in such a manner that 2/3rds of their basal portion is buried in the soil. Cuttings taken from main shoot or from the basal suckers root easily. Best results are obtained when basal portion of cuttings are treated with Keradix hormonal powder and planted so that 4-6 nodes are buried in the soil. Cuttings should not be obtained from those shoots which are less than 6 months old or more than 2 years old. Leaves are to be removed from cuttings before planting. In most parts of South India cuttings are planted during monsoon season and in Central and Western India they are planted during monsoon rains. Root initiation occurs within a fortnight after planting and will be ready for potting in 9 months. One-year-old rooted cuttings are best for planting. Pomegranate can also be propagated by ground layering, air layering or root suckers.

MANURE AND FERTILISER MANAGEMENT

Even though pomegranate grows well in soils of low fertility the producitivity can be increased greatly by adequate application of manures and fertilizers. If these are analysed best results in terms of yield and quality could be achieved. The doses of these largely depend upon soil type, age of the trees and crop load and also time of application.

For young plants, 2-3 months after planting 200 gm of Neem Cake along with 3-4 kg of well rotten FYM has to be applied for initial good start. Three months after this, application of 250 gms DAP, 250 gm Neem Cake along with 5 kg FYM is recommended. Again after 9 months each plant is to be applied with 500 gm DAP, 250 gm sulphate of potash, 1 kg neem cake and 10 kg well rotten FYM.

The following doses of fertilizers have been recommended

	Dose(g/tree)		
Age(years)	N	P205	K20
1½	250	125	125
2½	500	125	125
3½	500	125	125
6 and above	625	250	250

Alongwith the above fertilizers, 10 kg of FYM is to be applied in first year and the quantity may be increased to about 30 kg from fifth year onwards. These manures and fertilizers are to be applied to at the time of Bahar treatment and at different stages of fruits development.

The above fertilizer schedule may be adopted in our state till the research data is made available on the fertilizer schedule.

When the fruit is developing water soluble fertilizer is given through fertigation based on leaf analysis and the size and quality of fruits can be improved as per the export needs.

In drought-prone areas minor nutrient deficiencies appear frequently and interfere with fruit quality. Zinc deficiency leads to reduction of size in young leaves and margin curling. To correct deficiency of zinc 1-2 sprays of zinc sulphate at 0.6 percent is recommended. Cracking of young fruits will occur even if the water management is improper, when there is boron deficiency. It can be corrected by spraying borax @ g/L at 15-20 days interval or alternatively by the soil application of borax @ 12.5 g per tree at the time of fertilizer application.

Concerning the nutritional deficiencies commercial formulations such as Macrolic @ 0.33% at the time of fruit growth followed by 1% spray is advisable 15-20 days after the first spray, it results in increased fruits size, yield and quality. Similarly composite nutrient sprays of 0.4% ferrous sulphate + 0.3% manganese sulphate+ 0.2% Boric acid + 0.3% zinc sulphate, if sprayed before flowering, during flowering and fruit setting, results in increased yield and quality of fruits. Use of bio fertilizers like Azatobactor reduces the requirement of nitrogen by 2/3rd.

SAMPLING TECHNIQUE FOR LEAF ANALYSIS

It has been suggested that 8th pair of leaf (from growing tip) from new growth in April for Ambebahar crop and in August for Mrigbahar crop has to be collected. It should be from a vegetative terminal. Further it is suggested that composite sample of 50 leaves has to be collected from North, South, East and West directions.

LEAF NUTRIENT STANDARDS

Leaf nutrient ranges vary for satisfactory performance of pomegranate cultivars. The optimum ranges for N, P and K are 1.15 to 2.5, 0.13 and 0.39 and 0.24 to 0.9 respectively. The variation in the range may be due to the age of the trees, besides other factors.

MAJOR NUTRIENTS

Nitrogen

Nitrogen in pomegranate helps in increasing juice percentage, rind percentage and tritrable acidity. Total soluble solids, TSS/acidity ratio and content are also influenced by nitrogen.

Deficiency Symptoms: The symptoms are first noticed in old leaves and later in younger ones. The leaves become uniformly chlorotic yellow with light midribs and veins. In extreme cases, necrotic spots develop on the chlorotic margins of the leaves. The overall growth of the plant is checked and few thin branches develop. The leaf size is reduced and defoliation occurs. The yield reduces considerable with small fruits (Wavhal, et al., 1993)

Phosphorus

Deficiency symptoms: The suppression is overall observed. The shoots and stems become thin, number of branches and leaves are reduced; leaves of middle and apical portion of tertiary and side shoots become chlorotic yellow and the basal older leaves remain dull green and lose luster. The chlorosis starts at margins and tips of the leaves and margins of young leaves roll (Wavhal, et al., 1985).

Potassium

Deficiency symptoms: The growth of plant gets reduced producing lanky shoots. The size and number of leaves are reduced. The leaves at the middle position on the shoot develop chlorotic symptoms on the margins with low patches on the lamina having rusty brown or black grey spots. The backward curling of leaves from tips to bases is seen (Wavhal, et al., 1985) (Courtesy Production Technology of Arid - Semi Arid Fruits MPKV, Rahuri, 1996).

Correction: Potassium fertilizers such as potassium sulphate or chloride can be used

SECONDARY NUTRIENTS

Calcium

Deficiency symptons: Calcium being a relatively immobile element, deficiency symptons appear in young leaves. Reducing in growth of meristamatic tissue in young leaves and growing tips is an early indication. Calcium deficient leaves later become malformed and sizes reduce. This often leads to leaf-tip burn. It also affects fruits which become small in size and early maturity take place.

UNIQUE BENEFITS OF CALCIUM NITRATE (CN)

- Calcium improves strenght of cell wall, promotes root and shoot growth.
- CN makes plant healthy and tolerant to fungal and bacterial diseases, water stress etc.
- CN correct calcium deficiencies/disorders most effectively.
- Hidden deficiency of calcium, resulting in poor quality of produce is a common problem in many high value crops. The only solution to prevent this problem is to apply water soluble calcium as CN.
- CN improves quality and yield of produce.
- CN reduces post-harvest losses, improves shelf life and offers better storage properties of vegetables, fruits and flowers.
- The nitrate in CN is non-acidifying, non-volatile and the preferred form for most crops.
- Presence of Nitrate in CN promotes calcium uptake and vice-versa.
- The water soluble calcium and nitrate in CN are readily available at the root zone for quick uptake resulting in fast growth response of plants.

- Presence of calcium is low in acidic soils and availability of calcium is poor in alkaline and saline soils. In all these soils, CN works better than other sources like CAN, limestone, dolomite and gypsum.
- CN improves soil structure resulting in better porosity, water infiltration, root development and growth.
- The only fertilizer that provides water soluble calcium is CN, whereas CAN does not provide water soluble calcium.
- CN is the best source for all seasons, more so for winter/ cool weather conditions.

FIELD QUALITY (For broadcast application)

- Nursery: Apply 10-15 g/m^2.
- Mainfield: Suitable for basal and top dressing.

To meet the Calcium need, apply CN at the rate equivalent to 20 to 25% of recommended rate of nitrogen. Apply balanced quantity of nitrogen through other sources.

SOLUTION QUALITY (For fertigation and folliar spray)

- Fertigation: Use it alone or with other fertilizers (except phosphate and sulphate)as per practice.

  Generally, 0.5 to 1.0 g/L is recommended.
- Foliar application: Spray 5 to 10 g/L, 2 to 3 times at fortnightly interval, during the peak developmental/ reproductive phases.

Magnesium

Deficiency symptoms: In pomegranate magnesium deficiency is very common because pomegranate is normally grown under heavy soils. Magnesium deficiency in pomegranate is chlorosis mainly due to loss of chlorophyll; leaves are stiff, brittle and twisted. Overall suppression of growth, interveinal chlorosis of basal leaves with green midribs and veins, necrosis of the chlorotic areas and inverted 'V' shape of basal green area of the chlorotic scorched leaf are observed.

Correction: Magnesium should be applied at the rate of 18-24 kg/ha while maintaining a favourable balance between K and Mg. Spraying with a 0.2% magnesium sulphate solution in summers or a 0.5% magnesium sulphate in autumn is generally recommended as a prophylactic measure.

Sulphur

Deficiency symptoms: Deficiencies are same as those for nitrogen. Reduction in leaf size and chlorosis of leaves are seen. There is a pronounced effect on chlorophyll content of the leaves with suppression of growth so that the plants are stunted with few branches and leaves, heavy defoliation occurs at early stages. The younger leaves become chlorotic, which progresses to older leaves. Necrotic spots develop on the margins of the chlorotic leaves.

Correction: Sulphur containing fertilizers (Ammonium sulphate, super phosphate, and sulphur containing potassium salts) can be included in the fertilizer schedule.

Micro-nutrients

Micro-nutrients are beneficial in increasing yield as well as improving certain physical characteristics of pomegranate fruits.

Zinc

Zinc has a definite role in increasing the yield of pomegranate.

Application: Zinc can be applied in the form of spray (0.5 per cent zinc sulphate) three times at monthly intervals. Soil application @ 45 grams zinc sulphate per tree is recommended.

Iron

It has a role in increasing the number and weight of fruits. It has also a role in decreasing peel percentage.

Application: 0.4 percent spray with ferrous sulphate three times at monthly intervals.

Manganese

Manganese nutrition helps to increase the quality of pomegranate fruits.

Application: The spray of 0.3 per cent manganese sulphate is recommended by Bambal et al, (1991).

Copper

Chlorosis is reportedly a sysmptom for copper deficiency in pomegranate.

Application: Copper sulphate (0.4%) foliar spray increases the fruit quality and reduces chlorosis.

Boron

Boron improves rind and fruit colour. It also helps in improving percentage of grain and reduction in percentage of peel.

Deficiency symptoms: In young fruits, cracking occurs.

Corrections: Soil applications as borax (10gms per plant) or foliar application (0.2% boric acid) are the usual practice of boron application for pomegranate.

UNIQUE FEATURES OF SOLUBOR

- Boron plays a vital role in cell elongation, root and shoot development.
- Boron is essential for growing tips (root and shoot,) pollen germination and pollen tube growth.
- Boron has a primary role in cell wall biosynthesis and plasma membrane integrity.
- Boron facilitates short and long distance transport of sugars/photosynthates.
- Boron deficiency is common in many vegetables, fruit

crops, cereals, cotton, oil seeds etc. Even hidden deficiency inhibits flowering, seed development, and fruit development, etc.

- Seeds from Boron deficient crops have low germination rate with high percentage of abnormal seedlings.
- Sulubor effectively controls Boron deficiency in all crops and soils.
- Boron supplied through Solubor has better foliar uptake and distribution within plant system than that through Borax and Boric acid.
- Soil availability of Boron decreases with increasing soil pH, Boron deficiency is common in calcareous soils. Solubor is the best choice in all situations.
- Solubor improves quality of fruits, vegetables, grains, and seeds.
- Solubor is a recognized "organic fertilizer" in USA.
- Superior physical, chemical and biological properties of Solubor are the key factors for its fine quality and high efficiency.

GENERAL RECOMMENDATION:

Spray Solution Concentration: 0.1 to 0.15% (1 to 1.5 g Solubor/L). Higher or lower concentration can also be used depending on the nature of crops and their requirements.

Number of sprays: 2 to 3 sprays at 15 to 20 days interval at flowering/fruit setting stage.

Compatibility: Compatible with most agro-chemicals.

Suitability for fertigation: High suitable, better than all other Boron sources.

Suitability for soil application: Suitable for direct application as well as through water or herbicide spray.

Molybdenum

Deficiency symptoms: The symptoms are first noticed in old leaves and later in younger ones. The leaves become uniformly chlorotic yellow with light green midrib and veins. In extreme cases necrotic spots are also seen.

Correction: 1 to 2 kg/ha of sodium molybdate (corresponding to 400 to 800 g/ha of molybedenum) or equivalent amounts of ammonium molybdate can be used.

On acid soils with serious molybdenum deficiency molybdenum fertilization may even have virtually no effect unless the soil is limed at the same time. 0.1% sodium molybdate solution can be used for spraying purpose.

EFFECT OF SPILT APPLICATION OF 'N' AND 'K' FERTILIZERS ON GROWTH AND FRUITING ON GANESH POMEGRANATE.

Sl. No.	Fertilizer dose	No. of splits	Time of application
1	400:200:200 of NPK g/ plant (RDF)	1 dose	1st week of March
2	400:200:200 of NPK g/ plant	2 (at an interval of 2 months)	1st split is applied in 1st week of March
			2nd split is applied in 1st week of May.
3	400:200:200 of NPK g/plant	3(at an interval of 1½ month)	1st split is applied in 1st week of March.
			2nd split is applied in 3rd week of April.
			3rd split is applied in 1st week of June.

Sl. No.	Fertilizer dose	No. of splits	Time of application
4	400:200:200 of NPK g/plant	4(at an interval of 1 month)	Ist split is applied in Ist week of March.
			2nd split is applied in 3rd week of April.
			3rd split is applied in Istweek of May.
			4th split is applied in Ist week of June

Conclusion

In the overall situation it may be seen that four split application of fertilizer dose of 400:200:200 g NPK per plant (P applied in one split) has resulted in recording higher yield with maximum average fruits weight, juice percentage, TSS, Sugar, Ascorbic acid as compared to lower number of splits or one dose.

PARAMETER	UNIT	STANDARDS
pH	-	6.00 – 8.00
Electrical conductivity	dS/m	0.00 – 1.00
Organic carbon	%	More than 1.00
Available N	ppm	100 – 200
Available P	ppm	50 – 100
Available K	ppm	500 – 800
Available Ca	ppm	500 – 2000
Available Mg	ppm	350 – 500
Available S	ppm	10 – 50
Available Fe	ppm	2.00 – 20.00
Available Mn	ppm	2.00 – 20.00
Available Zn	ppm	1.00 – 5.00
Available Cu	ppm	0.20 – 2.00

ELEMENT	UNIT	STATUS
Nitrogen	%	2.94
Phosphorus	%	0.11
Potassium	%	0.76
Calcium	%	2.54
Magnesium	%	0.55
Sulphur	%	0.24
Iron	ppm	100
Manganese	ppm	44
Zinc	ppm	50
Copper	ppm	22

Leaf Nutrient Standards for Pomegranate

Nutrient	Observed	DRIS NORMS				
	Range	Defi-cient	Low	Opti-mum	High	Excess
N(%)	0.40–2.20	<0.54	0.54 - 0.90	0.91 - 1.66	1.67 - 2.04	> 2.04
P(%)	0.08–.33	<0.09	0.09– 0.11	0.12– 0.18	0.19-021	>0.21
K(%)	0.20–2.05	<0.20	0.20 - 060	0.61– 1.59	1.60– 2.26	>2.26
Ca(%)	0.06 – 2.40	<0.13	0.14 –0.76	0.77– 2.02	2,03– 2.65	>2.65
Mg(%)	0.16 – 0.49	<0.03	0.03 –0.15	0.16– 0.42	0.43– 0.55	>0.55
S(%)	0.04 – 0.70	<0.10	0.10 - 0.15	0.16– 0.26	0.26– 0.42	>0.42
Fe(ppm)	25 – 297	<34	34 – 70	71 – 214	215 – 286	>286
Mn (9ppm)	14 – 99	<15	15 – 28	29 – 89	90 – 119	>119
Zn (ppm)	7 – 44	<8	8 – 13	14 – 72	73 – 94	>94
Cu (ppm)	21 – 86	<7	8 – 28	29 – 72	73 – 94	>94
Yield t ha^{-1}	87 – 20	<13.7	13.7 –15.5	15.6 – 18.8	18.9 – 20.6	>20.6

1. Fertigation is a highly efficient technology which delivers fertilizers via irrigation systems, ensuring maximum efficiency of both water and dissolved nutrients usage.
2. These fertilizers are completely soluble in water.
3. The solution does not clog emitters from tubes pipes and other irrigation system.
4. These are low in salt index and hence do not leach or get fixed in the soil or increase salinity.
5. These are available in ionic form in water solution and hence are absorbed by the plant roots.
6. These contributes acidity to water which counter carbonate reactions, thus avoiding fixation of phosphorous with soil calcium.
7. These are chloride-free thus keeping salinity low and controlling negative effect of chlorides on plant metabolism.
8. As these fertilizers contains sulphates, they enhance incorporation of sugar and starch into fruits and storage organs, thus improving product quality.
9. The ratio of different nutrients in these fertilizers grade is such that it enhances uptake efficiency of each nutrient as it creates acidic soil solution in the root zone (pH 6-7).
10. Optimum plant nutrition demand of potassium and minimum demand of micro-nutrients is met by these fertilizers.
11. These are free flowing and easy to handle.
12. Every grain contains same ratio of nutrient as specified.
13. Split applications through every irrigation is possible without increasing labour cost of repeat fertilization as it occurs in conventional fertilizer application.

14. As these fertlizers are highly concentrated and losses on use are minimum the farmer need not carry large inventory. Also, smaller packages of 100 gms to 5 kg. besides 25 kg normal bags are available thus reducing inventory costs.
15. Nitrogen is largely available as Ammonium ion thus avoiding nitrate and amide losses. The ammonium ion easily gets converted to available form to plants as per its requirement and is not lost due to evaporation or leaching.
16. In calcareous soils, the continuous supply of P_2O_5 in low quantities minimizes fixation process of P.
17. The micro-nutrients are supplied in fully chelated form thus making micro-nutrients easily available to the plants in slightly acidic pH.
18. Electric conductivity of solution is low thus facilitating ionization and absorption by plant roots.
19. Water soluble fertilizer mixtures are used in low concentration. A stock solution of 10% is sufficient which can further be diluted to 100 times. Thus damage to plants due to high concentration of any nutrient is avoided.
20. Fixation, antagonizing, leaching and evaporation losses during application of these fertilizers is minimum. Thus these are ecologically friendly.
21. No harmful effect on micro-fauna and flora due to pH of 6 to 6.5. For horticulture and floriculture under green houses soluble fertilizers can be used and are most effective for high yield and high quality of produce. More and more farmers and cultivators are switching over to use of water soluble fertilizers specially in Maharashtra, Karnataka, Andhra Pradesh and Tamil Nadu. Nearly 330,000 hectares of land has come under drip/sprinkler irrigation systems. Water soluble fertilizer mixture requirement calls for nearly 80,000 tons per annum to meet the existing drip/sprinkler irrigation area.

FERTIGATION

A WHAT IS FERTIGATION?

- An obvious objective of every grower is to enlarge the income from the land we cultivates. The way to achieve this increase crop yield and minimize input costs.
- Among the various factors responsible for high crop yield, the use of proper quantity of fertilizers at appropriate time plays a vital role in enhancing the productivity.
- Fertilizers are normally used as basal dose for top dressing. The full year dose is split into one, two or three doses. Since these fertilizers are applied in bulk, lot of fertlizers go waste due to leaching, volatilization and fixation in the soil. Moreover these fertlizers get transmitted to areas beyond the active root zone and are not longer useful to the plant. In many cases, the effective utilization by the plant is less than 50% of the fertilizer applied.
- Drip irrigation is the most efficient method of irrigation for crops like fruits, flowers and vegetables. This system offers an opportunity for precise application of water soluble fertlizers and other nutrients to soil at the appropriate time with the desired concentration. This application of fertlizers through irrigation water is called fertigation.
- The development of root system is extensive in a restricted volume of soil when cultivation is done with drip irrigation and application of fertilizer or any other chemical through drip can efficiently place plant nutrients in the zone of highest root concentration. The major advantage of fertigation is that the necessary fertilizers can be applied uniformly to each and every plant even on a daily basis, thereby creating ideal and optimum environment for the plants to absorb required fertilizers and micro-nutrients.

FERTIGATION IN POMEGRANATE

Days	Grade	Quantity
1–15	12:61:0	1kg/day/AC
16–30	11:42:11	1 kg/day/AC
31–60	19:19:19	1 kg/day/AC
	+Urea	1 kg/day/AC
61–90	15:15:30	1 kg/day/AC
	+Urea	1kg/day/AC
91:120	0:52:34	1.5kg/day/AC
	+Urea	1kg/day/AC
	Next day calcium nitrate	1kg/day/AC
121-150	0:0:50	2kg/day/AC
	Next day calcium nitrate	1 kg/day/AC

B. WHY FERTIGATION ?

- Fertilizers can be applied timely according to crop requirements at different growth stages.
- Small quantities can be applied at close intervals.
- Irrigation combined with fertilizing maintains a solution in the root zone. The nutrients are readily available and uptake by the plants is fast and efficient.
- Every plant being irrigated is sure to receive its portion of nutrients.
- Even distribution of nutrients as well as water is maintained in the active root zone.
- Fertigation satisfies the requirements of low volume irrigation (micro sprinkles and dripppers) where root

system has a limited volume and hence accurate control of water and nutrients in the root zone is essential.

C. ADVANTAGES OF FERTIGATION

- Improves Availability of nutrient and consequently improves nutrient uptake by roots.
- Application is restricted to wetted area where root activity is most intensive.
- Loss of nutrients by leaching is minimized.
- Soil compacting is prevented.
- Weed population is reduced
- Reduces labour costs
- Lowers the fertilizers for the water use, hence saving in input cost.

D. METHODS OF FERTIGATION

Several techniques are available for applying fertilizers through drip irrigation systems. They can be classified into two groups according to their principle of operation.

a) SUCTION SYSTEMS:

1) **Ventury Type:** This is based on the principle of ventury tube. It delivers the fertilizers at a constant concentration which depends on the water flow. The ventury should be connected in parallel with the pipeline.

2) **Fertilizer Pump:** Fertilizer solution is prepared in tank from which it is pumped and injected into the irrigation system. The fertilizer solution is delivered at a constant concentration. The impeller and casing of such a pump (dozer pump) must be made of non-corrosive material, such as nylon or polycarbonate. The advantage of using a pump is that it builds up an additional pressure in the system unlike a ventury or a fertilizer tank where pressure drop takes place.

b) DELIVERY SYSTEM:

1) **Fertilizer Tank:** A tank containing fertilizer solution is connected to the irrigation pipe at the supply point. Part of the irrigation water is diverted through the tank, diluting the fertilizer solution and returning to the main supply pipe. The concentration of fertilizers in the tank thus becomes gradually reduced with time.

E. FERTILIZERS AND MICRO-NUTRIENTS:

1. NITROGEN: Nitrogen is most often applied through irrigation systems without any significant clogging problems. Careful consideration must be made of the pH in the irrigation water since some Nitrogen sources particularly aqueous ammonia and anhydrous ammonia will increase pH. Increased pH can result in precipitation of insoluble calcium and magnesium carbonates that can clog a drop system. Ammonium sulphate should not be used in places where calcium content of water is higher than 70 mg/lit.

2. **PHOSPHOROUS:** If we have to apply phosphorous individually through irrigation system, food grade (white) phosphoric acid is recommended. Green phosphoric acid, which is most commonly available and used in most of the complex fertlizers manufacturing, contains impurities and precipitates calcium. Maintaining low pH will help in controlling phosphate precipitation.

3. POTASSIUM: Potassium is also easily leached in sandy soils and generally must be replenished to maintain a proper N:K ratio for good crop production and quality. 'K' can be injected into the drip irrigation system as potassium chloride or potassium nitrate. These Potassium sources are soluble and have few precipitation problems.

4. MICRO-NUTRIENTS: Generally micro-nutrients can be applied pre-plant for most crops and in most soils. If supplied via the drip system, micro-nutrients such as iron,

zinc, copper and manganese can be injected as chlorides or sulphates. Sulphate salts of micro-nutrients may react with salts in the irrigation water and cause precipitation of clogging. Chelated micro-nutrients are highly water soluble and usually cause little clogging or precipitation.

F. ADVANTAGES OF USING 100% WATER SOLUBLE FERTILIZERS IN FERTIGATION:

- It ensures a regular flow of water nutrients resulting in increased growth rates and higher yield, yield increase of more than 100% is possible to achieve with correct fertigation schedule.
- These fertilizers offer greater versatility in the timing of the nutrient application to meet specific crop demands. Fertilizers can be applied at pre-determined times according to the developmental and physiological stages of the crop. It should be noted that delivering the fertilizer in split doses rather that in a single application results in a more pronounced increase in the plant nutrient uptake.
- It improves the availability of nutrients and their uptake by roots.
- The three major nutrients N, P and K, supplied in one solution to the plant, gives root absorptions and consequently higher yields.
- Safer application source which, eliminates the danger of burning the plants, root system, since these fertilizers are greatly diluted in irrigation water.
- Offers simpler and more convenient application than any soil application of fertilizers, thus saving time, labour, equipment and energy.
- Insecticides and herbicides can also be supplied along with these fertilizers through irrigation system since they are compatible with most of them.

G. CATEGORIES OF WATER SOLUBLE FERTILIZERS:

The fertilizers can be classified into two categories basing on the nutrient sources chosen for manufacturing them.

1) **Chloride free fertilizers:** These fertilizers are useful for high value crops and crops which are more sensitive to chlorides such as tobacco, certain fruits such as citrus, grapes, pomegranate and vegetables such as Tomato, Potato, cauliflower etc. These fertilizers are produced by using urea, ammonium nitrate, white phosphoric acid and potassium and nitrate as basic ingredients. There are a very few (2 or 3) companies in the world which are supplying this type of high quality fertilizers e.g., Haifa Chemicals Ltd, Haifa, Israel.

2) **Normal fertilizers:** These fertilizers are produced by using ammonium nitrate, urea, ammonium Phosphate, Ammonium Sulphate, green phosphoric acid, potassium, chloride and potassium sulphate. Since these fertilizers contain chlorides and sulphates, there is always possibility of clogging of irrigation system due to precipitation of calcium and magnesium bicarbonates, sodium etc, if irrigation water pH is not maintained. Presence of high concentration of chlorides in the wet zone will lead to more uptake of chlorine by plants ultimately leading to shorter shelf life and bad quality of fruits, vegetables, flowers etc. Most of the foreign companies, who have recently entered into Indian markets, are supplying this category of fertilizers.

H. LIST OF FERTILIZERS NOT TO BE USED FOR FERTILGATION:

1. Aqueous Ammonia (NH_4OH)
2. Calcium Nitrate
3. Calcium Ammonium Nitrate
4. Potassium Sulphate

5. Zinc Nitrate
6. Ferric Sulphate

1. SUITABLE FERTILIZER AND NUTRIENT CONCENTRATION FOR FERTIGATION:

1. **FERTILIZER CONCENTRATION:** The concentration of fertilizers in irrigation water can range from 4000 to 10000 ppm. Normally 0.5% concentration (5000 ppm) is maintained for drip irrigation system which is equivalent to 5 gm/litre water. The actual concentration needed depends on the fertilizing material and crop requirement.
2. **NUTRIENT CONCENTRATION:** For proper growth, plant must be supplied with nitrogen, phosphorous, potassium, calcium, magnesium, boron, zinc, copper, magnanese, molybdenum, Iron etc. The following contains the nutrient concentrations which seem best suited for a wide range of crops:

ELEMENT	CONCENTRATION
Nitrate Nitrogen (NO_2)	150-200 PPm
Phosphorous(P_2O_5)	110 PPm
Potassium (K2O)	240-490 PPm
Calcium (Ca)	150-22 PPm
Magnesium(Mg)	20 PPmm
Iron(Fe)	5 PPm
Manganese(Mn)	1 PPm
Copper(Cu)	0.1 PPm
Zinc(Zn)	0.1 PPm
Boron (B)	0.2 PPm
Molybdenum(Mo)	0.05 PPm

Zn – EDTA – 15%

This product is already covered in FCO and it is widely used in India. This is a free flowing crystalline powder with 15% zinc. In this product the zinc is chelated by a chelating agent EDTA (Ethylene diamante tetra acetic acid). The pH of the product is 6 to 6.5 with particle size 95% <0.5 mm. This product is recommended for soil application as well as for foliar application.

The zinc deficiencies are always reported on horticultural crops like pomegranate, grapes, mango and other horticultural crops. The Zinc EDTA through soil application as well as through foliar application is useful to correct or prevent the zinc deficiencies. It is useful for uniform sprouting as well as uniform vegetative growth. This product is very well suited for soils ranging from pH 5.0 to 10.0.

Zn EDTA contains 15% Zinc in which there is 15% and other chelating agent and soluble filler materials.

Fe – EDTA – 13% Fe

Fe – EDTA 13% Fe and the other are chelating agents. Ethylene diamante tetra acetic acid and other soluble fillers materials. In India, this product is widely used to correct or to prevent the iron deficiencies and iron chlorosis and it is recommended for soil pH ranging from 4 to 8.

This is a free flowing homogeneous product with pH 5.5 to 6.5. The Fe with the chelating agent is readily available to the crops when given through spraying.

Fe (EDDHA) – Fe 6%

This is the source of Iron. Iron is a micro-nutrient which is required for plant growth and without disorders like iron chlorosis it results in reduction in yields. Earlier, this nutrient was supplied through product like chelated Iron as Fe–EDTA. But for soil application the chelating agent like EDDHA proves better than EDTA.

EDDHA is a chelating agent which is a short form of Ethyleedianin-di-(0-hydroxyphenyl acetic acid). Iron is chelated with the chelating agent EDDHA. Fe (EDDHA) contains 6% Fe with pH value 7-8. This is a homogeneous, micro-granulated free flowing product with particle size 95%<0.5mm.

Fe EDDHA is recommended for soil application of micro nutrients either through drip or through drenching. For preventive treatment the lower application rates are sufficient while to correct the deficiencies, the higher rates are recommended.

Fe (EDTA), Fe –12% is a product which comes under FCO which is a chelating agent, ethylene damien tetra acetic acid. This product is recommended only for the soil pH ranging upto 8. This product is not suitable for the pH range more than 8 and above. Considering the limitation in limited pH range recently new chelating agent is invented as EDDHA. This product is recommended on wide pH ranging from 4.0 to 12.0. The applications rates are:

	Gm/plants	
	Initial production	Full production
Horticultural crops	40-60	75-100
Grapes	10-15	15-70
Grapes (wine purpose)	5-7	7-10
Tomato, brinjal etc.	60-80 gm/gunta	
Strawberries	100-150 gm/gunta	
Melons	100-150gm/gunta	

Fe EDDHA contains 6% Fe and rest of the materials are chelating agent and other soluble fillers.

The Water Soluble fertilizer has the following merits over the conventional fertilizers:

1. Soluble fertilizers completely dissolve in water leaving no precipitation. Therefore, there is absolutely no problem

in fertigation, clogging of drippers and emitting pipes.

2. Every plant irrigated receives a regular flow of both water and nutrients directly in the root zone resulting in improving growth and increasing the productivity. Over two-fold increase in yield is possible due to correct fertigation schedule.

3. Nutrients can be applied as and when required to meet specific crop nutrient demands. Also fertilizers can be applied at a predetermined time according to the critical growth stages of crops. Frequent fertilizers (more split doses)versus single application in conventional fertilizers results in remarkable plant nutrient uptake which refflects in increased bio-mass production.

4. Since the nutrients directly reach the rhizosphere in liquid form there is no loss of nutrient especially through volatilization.

5. The fact that three major and micro-nutrients are supplied in one solution to plants, it gives better root absorption and consequently higher yields and good quality.

6. Safer application methods eliminate the danger of burning the plant root system as the fertilizer is greatly diluted in the irrigation water.

7. Since these fertilizers can be applied though drip irrigation through venture, fertilizer tank or injection pump, the application becomes very simple than any soil application of fertilizers and thus results in saving labour, time, energy and overall application cost.

8. Due to frequent application in small doses, the amount of fertilizers present in the soil at any time is small which prevents losses from leaching and run off during heavy rainfall. This results in saving of nutrient requirements per unit area which ultimately results in saving of fertilizers nearly by 25 to 50 % of the recommended level.

9. Wide range of compatibility of most soluble fertilizer grades facilitates concurrent incorporation of fungicides, pesticides and nematicides, which helps saving in labour, time, energy and its application cost.

10. Since most soluble fertilizers are blended with chelated micro-nutrients, there is no necessity to go for additional micro-nutrient mixture or sprays which ultimately helps reduce the additional cost on labour and mixtures.

11. Allows crops to be grown on marginal lands, such as sand or rocky soils, where accurate control of water and fertilizers in the plant's root is critical.

Water Soluble Fertilizers

N:P:K

1:48:35

Technical Aspects

Analysis:

Total nitrogen N	:	1%
Nitrate, NO3N	:	1%
Ammonium NH4N	:	0%
Phosphorus (P_2O_5)	:	48%
Potash (K_2O)	:	35%

Agronomic Benefit

100% water-soluble fertilizer in a crystalline form; The NPK is especially developed to meet with all the nutrient requirements of the crops during the second stage after planting.

Increase of yield and better fruit quality. The efficient root uptake of nitrogen, phosphorus and potassium through FERTIGATION guarantees an optimum metabolism stand growth of plants and fruits (formation of nucleic acids, flowering, photosynthesis, perspiration, formation and transport of sugars).

Fertigation

In continuous fertigation, the concentration of the nutrient solution should be between 0.5-1.5 g/I/For the mother solution, a concentration of 10-15% can be used depending on the water temperature. The number of fertigation applications should be spread in the corresponding period of application.

There are several differences between the fertilization of crops grown in greenhouses and those grown in open fields. The uptake of nutrients is higher in protected crops due to faster growth and higher yield. The nutrient requirements also depend on climate conditions, soil type, temperature, varieties, methods of application and irrigation.

Use: *Prepare a mother solution by adding 15 kg NPK 1:48:35 to 1000L water (20°C). This solution contains:

1.5 gr. N/litres solution

72.0 gr. P_2O_5/litres solution

52.5 gr. K_2O/liters solution

Water Soluble Fertilizers

N:P:K + TE

19:19:19 + TE

Technical Aspects

Analysis:

N	:	19%
P_2O_5	:	19%
K_2O	:	19%

Trace Elements

Mgo	:	1%

B	:	0.01%
Mo	:	0.001%
Cu	:	0.01%
Mn	:	0.02%
Zn	:	0.02%
Fe	:	0.04%

Agronomic benefits

100% water-soluble fertilizer in a crystalline form; The NPK is especially developed to meet all the nutrient requirements of the crops, during the second stage after planting.

The metallic trace elements are in chelated form in which the trace elements are protected against fixation in the soil and remain available for the plants. It contains no chlorine or sodium and is most suitable for the intensive crop production, and for the nutrition of saline-sensitive crops.

There are several differences between the fertilization of crops grown in greenhouses and those grown in fields. The uptake of nutrients is higher in protected crops due to faster growth and higher yield. The same applies to crops grown in spring and summer, verses crops in winter. The nutrient requirements also depend on climate conditions, soil type, temperature, varieties, methods of application and irrigation.

Use: *Prepare a mother solution by adding 15 kg NPK 1:48:35 to 1000L water (20^{0}C). This solution then contains:

28.5 gr. N/litres solution

28.5 gr. P_2O_5/litres solution

28.5 gr. K_2O/litres solution

1.5 g. MgO and a well-balanced micro-nutrient mix is developed to provide the crops with all the necessary trace elements in the right proportion.

Water soluble Fertilizers

N:P:K + TE

20:10:10+TE

Technical Aspects

Analysis

N	:	20%
P_2O_5	:	10%
K_2O	:	10%

Trace Elements

Mgo	:	1%
B	:	0.01%
Mo	:	0.001%
Cu	:	0.01%
Mn	:	0.02%
Zn	:	0.02%
Fe	:	0.04%

Agronomic Benefits

100% water-soluble fertilizer in a crystalline form; The NPK is especially developed to meet all nutrient requirements of the crops during the second stage after planting.

The metallic trace elements are in chelated form in which the trace elements are protected against fixation in the soil and remain available for the plants. It contains no chlorine or sodium and is most suitable for use in intensive crop production and for the nutrition of saline-sensitive crops. As it contains no urea, the nitrogen in this NPK acts very rapidly.

There are several differences between the fertilization of crops grown in greenhouses and those grown in open fields. The uptake of nutrients is higher in protected crops for faster growth and higher yield. The same applies to crops grown in spring and summer, versus crops grown in winter. The nutrient requirements also depend on climate conditions, soil type, temperature, varieties, methods of application and irrigation.

Use:* Prepare a mother solution by adding 15 kg NPK 20:10:10 + TE to 1000L water (20°C). This solution then contains:

37 gr. N/litres solution

15 gr. P_2O_5/litres solution

15 gr. K_2O/litres solution

1.5 gr. MgO and a well-balanced micro-nutrient mix is developed to provide the crops with all the necessary trace elements in the right proportion.

*By applying 66,71 of the mother solution, you supply the crops with 2.5 kg N, 1 kg P_2O_5, 1 kg K_2O, 0.093 kg MgO and all the trace elements.

Water Soluble Fertilizers

N:K

13 : 0 : 45

Potassium Nitrate

Technical Aspects

100% macro-nutrient content

Potassium K_2O (available as K^+)	45%
Nitrogen N (available as NO_3)	13%
Completely soluble in water with insoluble	<0-0.75%

Content

Solubility at 20°C : 315 gram PK/liter of water

Free of Chlorine, Sodium and heavy metals

Chlorine (Cl^-)	max 0.5%

Sodium (Na^+)	max 0.5%
Heavy metals	<10 ppm

Easy to mix, PK is compatible with all water soluble fertilizers

pH at 1% solution in water 8-9

Low salt index and low EC value, 1 g/l at 25°C 1.3 mS/cm

Potassium nitrate's key advantages

High purity fertilizer	:	Potassium nitrate contains no elements detrimental to plants: it is free of chlorine compounds, Sodium and heavy metals
Fully soluble	:	Potassium nitrate is a free-flowing fine crystalline powder that dissolves quickly in water.
Compatibility	:	Potassium nitrate can be mixed with all water-soluble fertilizers, Potassium nitrate is compatible with the majority of pesticides in foliar application
Nitrate nitrogen	:	Nitrate nitrogen is non-volatile and enhances the uptake of (Ca^{++}, Mg^{++}.. ..). Nitrate nitrogen is the most efficient source for plant growth.

No salt accumulation	:	Potassium nitrogen can be used to cover the potassium needs of a crop without supplying excess of Sulfate or Chlorine.
Allround	:	Due to its low N/K ratio, Potassium nitrate is suitable for all crops and growing stages, including flowering and ripening stages.

Fertigation

Potassium Nitrate is an indispensable components in all well-balanced fertigation programmes, both for soil-grown crops and hydroponics' systems. Potassium nitrate, used as a K-source, will contribute part of the Nitrate nitrogen. It is the only K-fertilizer suitable for intensive cropping without risk of sulfate or chlorine salt accumulation. Under saline or alkaline conditions, potassium nitrate is the only suitable K-choice.

For intensive horticulture, nitrate nitrogen is the optimum Nitrogen source. A high NO_3/NH_4^+ balance assures sufficient uptake Ca^{++} and Mg^{++} and avoids root degeneration, especially at high soil temperatures. Additionally Nitrate Nitrogen losses by volatilization are avoilded.

For hydroponics' systems, the most intensive and sensitive technique, this is even more important. KNO_3 in a hydroponics' system is imperative and today's horticulture would not be the same without potassium nitrate.

Foliar Application

Due to low risk of leaf burning (low EC-value osmotic pressure) and wide compatibility with other fertilizers and pesticides, potassium nitrate is a recommended foliar fertilizer.

Foliar application of potassium nitrate is an excellent way to supply additional nitrogen and potassium during critical physiological or environmental conditions. Furthermore, important physiological processes such as dormancy break (mango and deciduous fruits), ripening fruit maturation, etc, can be influenced with foliar applications of potassium nitrate.

Water Soluble Fertilizers

NP

12:61:00

(Mono Ammonium Phosphate)

Technical Aspects

100% macronutrient content

Nitrogen (NH_4)	12%
Phosphorus (P_2O_5)	61%

Completely soluble in water with a low insoluble content <0.01%

Solubility at 20° C :365 g/litres

Free of Chlorine (Cl^-) and heavy metals

Chlorine (Cl^-)	5-10ppm
Heavy metals	<10 ppm
Easy handling, free flowing	0.05%
Humidity	
pH at 1% solution in water	4.5

Low salt index and low EC value

Salt index	30%
E.C value 1 g/l at 25° C	0.86 ms/cm

MAP (Mono Ammonium Phosphate) is a fully water-soluble fertilizer, composed of 100% plant nutrients with 84%

di-hydrogen-phosphate ($H_2PO_4^-$) and 17% ammonia (NH_4^+). MAP is therefore the water-soluble fertilizer with the highest P- content available. This, combined with its low ammonia content, makes MAP suitable for fertigation in soil-grown crops.

Agronomic Benefits

Used mainly as a P-source, MAP is the perfect product to supply phosphate and nitrogen simultaneously in fertigation programs for soil-grown crops

In intensive fertigation, control over N availability is limited if urea is used as nitrogen source. MAP's low ammonia content maintains the correct nitrogen balance between ammonia and nitrate, avoiding the use of urea. Combination with any nitrate nitrogen source ensures an optimum NO_3/NH_4^+ ratio for any crop.

MAP's key Advantages

High-purity fertilizer	:	MAP contains no elements detrimental to plants: it is free
Fully soluble	:	MAP is a free-flowing crystalline powder that dissolves quickly in water.
Compatibility	:	MAP can be mixed with all water-soluble fertilizers other than calcium-containing fertilizers.
pH Buffer	:	the Phosphate in MAP buffers fertigation solutions, keeping the pH effectively stable at around 4.5.
Urea-free	:	Nitrogen in MAP is present in the ammonia

form, offering better control over N-availability to plants than urea.

Water soluable Fertilizer

K

00:00:51 + 18% S

(Potassium Sulphate)

Technical Aspects

100% macro-nutrient content	
Potassium K_2O	51%
Contains additional essential sulfer (S-SO_3)	18%
Solubility at 20° C : 111 gram K/litres.	
Free of Chlorine (Cl^-), sodium (Na^+) and heavy metals	
Chlorine (Cl^-)	1% max
Sodium (Na^+)	1% max
Heavy metals	<10 ppm
Easy handling, free flowing	
Humidity	0.5%
pH at 1% solution in water	3.0
Low salt index and low EC value	
Salt index	30%
E.C. value 1g/l at 25°C	

Potassium sulphate is a water-soluble fertilizer, composed of 100% plant nutrients with 44% potassium (K_+) and 56% sulphate (SO_{4-}). Due to the absence of Nitrogen and the high potassium content, potassium sulphate is a useful fertilizer for drip irrigation, especially in areas with sufficient water availability.

Agronomic Benefits

In fertigation, accurate control of the supply of every nutrient is a key to nutrition management. Potassium sulphate is a chlorine-free, water-soluble fertilizer.... the only potassium fertilizer that also provides the crop with sulfur, an essential secondary macro-nutrient. As it is free of other macro-nutrients, potassium sulphate is a useful too that allows the application of potassium at the required levels at any growing stage. Potassium sulphate is an excellent fertilizer for providing part of the potassium requirement, being careful not to exceed the sulphate needs of the plant. Thus avoiding excess use of it.

It can be combined with any Nitrogen source according to crop requirement, controlling the K/N ratio as desired.

In drip irrigation systems on alkaline soils, the low pH of potassium sulphate contributes to a slight acidification of the zone surrounding the roots (rhizosphere), enhancing the availability of phosphate and micro-nutrients present in the soil.

In hydroponics, potassium sulphate is used as a sulphate source for crops that demand high sulfur.

Potassium Sulphate's key Advantages

High Purity Fertilizer	:	Potassium sulphate contains no elements detrimental to plants: it is free of chlorine compounds, sodium and heavy metals.
Fully soluble	:	Potassium sulphate is a free-flowing, fine crystalline powder, soluble in water.
Compatibility	:	Potassium sulphate can be

		mixed with all water-soluble fertilizers, other than calcium-containing fertilizers.
Flexibility	:	The high concentration of potassium and absence of nitrogen and phosphate enables the Correct quantity of potassiunm to be supplied at any stage. Potassium sulphate can be combined with N, P or NP-fertilizers according to crop needs.
Low pH	:	In alkaline soils, potassium sulphate lowers the pH in the rhizosphere, helping to improved availability of phosphates and certain trace elements. In drip irrigation solutions it helps to reduce the pH of the irrigation water.

NPK 19-19-19 + 1.5 MgO + TE

Specifications:

Nitrogen (N)	19.0
Ammonium, NH_4-N	3.7
Nitrate, NO_3-N	5.6
Urea, NH_2-N	9.7
Phosphorous (P_2O_5)	19.0
Potassium (2O)	19.0
Magnessium (MgO)	1.5

Sulphur (S)	1.2
Trace Elements:	
Iron (Fe)EDTA- chelated	0.04
Manganese (Mn) EDTA chelated	0.02
Zinc (Zn) EDTA chelated	0.02
Copper (Cu) EDTA chelated	001
Boron (B)	0.01
Molybdenum (Mo)	0.001
EC value 1 gram/litres at 25°C (ms/cm)	

Water Soluble Fertilizers

P:K

0:52:34

Technical Aspects

100% macro-nutrient content	
Phosphor P_2O_5 (available as H_2PO_4-)	52%
Potassium K_2O (available as K_+)	34%
Completely soluble in water with a low insoluble content	< 0.1%
Solubility at 21°C: 180 gram PK/litres.	
Free of Chlorine (Cl^-), sodium (Na^+) and heavy metals	
Chlorine (Cl^-)	20-60ppm
Sodium (Na^+)	<10 ppm
Heavy metals	< 10 ppm
Easy handling, free flowing no caking	
Humidity	0.2-0.5%

pH at 1% solution in water

Low salt index and low EC value

Salt index

Agronomic Benefits

1. Increase of yield and better fruit quality. The efficient root uptake of both phosphate and potassium of PK in FERTIGATION guarantees an optimum metabolism and growth of plants and fruits (formation of nucleic acids, flowering, photosynthesis, respiration, formation and transport of sugars.....)

2. The physico-chemical characteristics of 0.52:34, pH, high solubility and low salt index makes the product particularly suitable for:

- Application as a foliar fertilizer (low risk for leaf burning and scorching)

- Application in fertirrigation as a regular P source, especially for irrigation of water with a high salt concentration since PK has a very low conductivity value (EC value).

DISEASES IN POMEGRANATE

POMEGRANATE (Punica granatum L.)

1. Cercospora leaf spot - Cercospora punicae
2. Bacterial blight - *Xanthomonas axonopodis* pv. *Punicae*
3. Leaf spots - *Colletotrichum gloesporioides*
   - *Sphaceloma punicae*
   - *Fusarium fusarioides*
   - *Phomopsis aucubicola*
   - *Drechslera rostrata*

4.	Minor diseases		
a.	Canker	-	Ceuthospora phyllositicia
b.	Leaf and fruit spot	-	Coelophoma empetri
c.	Flower and fruit spot	-	Phytophthora nicotianae
d.	Fruit spots	-	Beltaraniellla humicolla
		-	Pestalotiopsis versicolor
e.	Fruit rots		
i.	Cladosporium fruit rot	-	Cladosporium oxysporum
ii.	Aspergillus fruit rots	-	Aspergillus spp.
iii.	Mild soft rot	-	Penicilliium Chrysogenum
iv.	Soft rots	-	Rhizopus arrhizus & Stolonifer
v.	Dry rot	-	Syncephalastrum racemosum
vi.	Fusarium rot	-	Fusarium equiseti
vii.	Phomopsis rot	-	Phomopsis sp.

1. Cercospora leaf spot

Symptoms: Light brown zonate spots appear on the leaves and fruits. Black and elliptic spots appear on the twigs. The affected areas in the twigs become flattened and depressed with raised edge. Such infected twigs dry up. In severe cases the whole plant dies.

Fungus: Cercospore punicae P. Henn. Conidiophores are olivaceous brown, short, fasciculate, sparingly septate and 20 to 40 x 3 um. Conidia are hyaline to pale olivalceous cylindric, sub-fusoid to sub-clavate, septate and 40 x 50 x 3.0 um.

Mode of spread and survival: The pathogen spreads through wind-borne conidia.

Epidemiology: The disease is serious during Sep-Nov.

Management: The disease can be effectively controlled by pruning and destruction of diseased twigs followed by spraying with thiophanate-methyl 0.1 per cent or chlorothalonil 0.2 per cent or mancozeb 0.2 per cent.

2. Bacterial blight

Symptoms: Small irregular, water-soaked spots appear on the leaves. Spots vary from two to five mm in diameter with necrotic centre of pin-head size. Spots are translucent, later turn light brown to dark brown and are surrounded by prominent water-soaked margins. Spots coalesce to form large patches. Several infected leaves fall off. The bacterium attacks stems, branches and fruits also. On the stem, the disease starts with brown to black spots around the nodes. It leads to girdling and cracking of nodes. Finally the branches break down. Brown to black spots appear on the pericarp with L or Y-shaped cracks. The spots on fruits are raised and are oily in appearance.

Bacterium: Xanthomonas axonopodis pv. Punicae (Syn. Xanthomonas campestris pv. Punicae. It is a Gram-negative rod, motile with single polar flagellum and 0.5 to 1.0 to 2.5 um. It is non-acid, fast and aerobic.

Mode of spread and survival: The bacterium survives on the tree. The pathogen survives for 120 days on the fallen leaves during off season. The primary infection is through infected cuttings. The disease spreads through wind splashed rains.

Epidemiology: High temperature and low humidity favour the disease.

Management: Clean cultivation and strict sanitation in the orchard help to reduce the disease incidence. Spraying with

Bordeaux mixture 1.0 per cent controls the disease.

- Copper oxychloride 3g + Bromopal 0.5 g/lit.
- Apply 20-25 kg Bleaching powder to ground
- Dip seeatures in sodium hypochlorite 25 ml / lit of water.
- Take Hastabahar (Sept - October) crop
- Spray Bordeux mixture 1% before pruning & after Ethrel application
- Collect the Leaves & bun.

3. Leaf spots

a. Colletotrichum gloeosporioides Penz: The disease appears as small with regular to irregular dull violet or black spots on the leaves. These spots are surrounded by yellow margins. The infected leaves turn yellow and drop off. The pathogen spreads through wind-borne conidia. The disease is severe during August-September when there is high humidity and the temperature between 20 and 27°C.

The disease is effectively controlled by spraying with carbendazim 0.1 per cent or thiophanate-methyl 0.1 per cent or mancozeb 0.2 per cent at fortnightly interval.

b. Sphaceloma pounicae Bitancourt & Jenkins: The disease attacks leaves, shoots, calyx and fruit. Rusty spots appear on leaves. Infected leaves turn yellow and die. Rusty coloured pustules appear on fruits. Drizzling rains and abundant dew favour disease development and spreads. Spraying with thiophanate methyl 0.1 per cent or carbendazim 0.1 per cent controls the disease.

c. Pestalotiopsis versicolor (Speg.) Stey: The disease manifests its symptoms as minute, brown to rust coloured spots on the fruits. The spots coalesce as the disease advances and causes necrotic patches. The central portion of the lesion is depressed inward with raised margin and severe infection tear open the rind. In severe cases infection penetrates deep into the fruits and causes discolouration of seeds. Mycelium of the fungus is cottony white with yellowish tinge with branches. Conidia are borne on short conidiophore. They are 5-celled, oblong to clavate, 21.8 x 4.3 um in dia. Intermediate three cells are dark coloured and long. Apical cells on either side are hyaline. Apical conidial cell bears 3 hyaline cells. The basal cell is short, obtuse, hyaline with a small pedicel.

d. Fruit rots:

i. Cladosporium oxysporum Berk & Curt: The diseased fruits develop orange-red to dull-brown circular spots and become olive-brown. In advanced stage, the entire fruit rots. Hyphae of the fungus are septate, light olive green, 2.5 to 3.0 um in width. Conidiophores are light brown and simple. Conidia are light brown to olive green, 1-celled fusoid and 1 to 20 x 3.5 to 4.5 pm.

ii. Aspergillus fruit rots: Aspergillus flavus Link; It causes brownish dicolouration, which gradually becomes blackish and slimy. Later, it gets slightly depressed and is covered by green conidial heads of the fungus. The disease causes soft rot of fruits and emits fermented odour. Aspergillus Niger: Symptoms first appear in the form of a brownish discolouration which gradually become blackish brown. In advanced stages, a major portion of the fruit is involved giving it a depressed and slimy appearnace. Rotten fruits emit a fermented odour. *Aspergillus niveus Bloch*: It produces small brownish patch which increases in dia. and turns darker with blackish tinge in the centre. The diseased tissue shrinks and ultimately disintegrates especially in the central region, followed by irregular depression and exudation of slimy mass. The diseased fruits emits foul odour. *Aspergillus versicolor* : It produces small brownish patch, increases in size and turns darker with blackish tinge in the centre. The diseased tissue shrinks and ultimately disintegrates especially in the central region followed by irregular depression and exudation of slimy mass emitting foul odour.

iii. Mild soft rot: Penicillium chrysogenum Thom. Soft, waterly sport of two to four cm dia appear on the fruits. These spots enlarge and coalesce together. The spots are found covered with white mycelium and bluish green spores.

iv. **Soft rot:** Rhizopus arrihizus Fisher and R. stolonifer Ehr. Ex Fr; Small spots appear on fruits. They increase in size and coalesce. Infection is restricted to rind. But the entire internal content decays into a pulpy mass. Under dry conditions cracking originates from the point of infection. Packing straw should be treated with sulphur dioxide. Treatment of fruits with linseed oil or mustard oil or castor oil protects them from rot by R. arrhizus.

vi **Dry rot:** Syncephalastrum racemosum Cohn: Small isolated dark patches are formed on the surface of the fruits. The pathes are dry and covered with the mycelium and spores of the fungus. Inner pulp rots.

vi. Fusarium equiseti (Corda) Sacc: Circular and depressed lesions appear on the fruits. Lesions are surrounded by concentric wrinkles. The lesions increase in size and cover almost the entire fruit.

vii. Phomopsis sp: The disease starts from calyx-end and gradually spread over the entire fruit. Pycnidia appear on affected areas. One spray with copper oxychloride checks the spread. Copper oxychloride 0.5% sprayed three times at 10 days interval, controls the disease.

e. Fusarium fusarioides (Frag & Cif.) Booth. The disease appears as minute spots towards the leaf margin. The spots are brown, circular to irregular in shape. Later the spots coalesce and form big dark brown necrotic blotch.

d. Phompsis aucubicola Grave: Buff-brown spots appear on the margins sometimes spots are found scattered on the leaves. Black pycnidia are found on the spot in the upper surface of the leaf. Hyphae of the fungus are colourless, poorly branched, globose to sub-globose, 201.6 to 486 pm in dia, dark brown to black and ostiolate. Conidiophores are simple, short, unbranched, hyaline and 4.6 x 8.2 pm long. Alpha spores are hyaline, ovoid to along or rearely fusoid to sub-fusoid, often biguttulate and 7.4 to 21.8 x 4.3 to 3.1 pm. Beta spores are filiform, curved or unicinate and measures 19.8 to 34.4 x 0.54 to 1.4 pm.

e. Drechslera rostrata (Drech.) Richardson & Fraser. The disease is characterized by the appearance of numerous, small, black spots scattered all over the fruit. The margin of the spots varies from dark green to orange in colour. In advanced stages, few spots gradually enlarge and coalesce to form big dark spots of various sizes. Mild infection is confined to rind of the fruits but severe infection extends to the inner tissues and even to the seeds, showing ashy discolouration. The fungus produces conidia which are-pale to dark olivaceous brown, cylindrical to rostrate, somewhat less curved, measuring 43.1 to 124.9x9.2 to 17.4 pm in size with 5 to 13 transverse septa and a minute hilum protruding from the base. The other end is bluntly tapered. The septa from basal and apical cells are prominent being thicker and darker than the intermediate septa.

4. Minor diseases

a. Canker: Ceuthospora phyllosticta C. Mass; Elluiptic, black spots are formed on the twigs. Affected areas become flattened and depressed with raised edge. Later the bark dries and cracks and the wood below shows abnormal dark brown-black discolouration. Twigs beyond the cankerous spots dry off and in severe cases the affected tree dies.

b. Leaf and fruit spot: Coelophoma empetri (Rostrup) Petra; On leaves the spots are circular, reddish brown to dark brown. They coalesce to form bigger sized lesions which are necrotic and dark brown. Later infected leaves turn pale yellow and drop down. On fruits numerous, minute, cirucular, tan brown spots which turn brown to black later. They coalesce to form irregular, depressed and hard necrotic lesion. Lesions are restricted to epidermis bearing black, spherical pycnida. Mycelium of the fungus is olive-grey when fresh and it turns dark violet at maturity. They are separate, branched, and 5.5, pm in width. Pycnidia are dark brown to black, spherical to irregularly elongate and 335.2 x 26.2 to 61.0 pm in dia. Pycnidiospores are one septate, elongate to cylindrical with round to pointed ends and 11.8 x 1.6 pm.

c. Flower and fruit spot: Phytophthora nicotianae Breada de Haan; Spots on flowers leads to premature shedding. Lesions are also found on fruits. Twigs in the trees are also infected. Hyphae of the fungus are 4.8 pm in dia with hyphal swellings. Chalamydospores are intercalary or terminal. Sporanging are broadly turbinate with spherical basal portions and apical part prolonged into a beak, papillate and 63 x 44 pm. Sporangia do not shed.

d. **Fruit spots:**

i. Belta raniella humicola: The disease is characterized by appearnace of black circular spots which gradually enlarge and coalesce to form big dark black spots leading to necrosis. The margin of spots vary from reddish to brown in colour. Infections restricted to the rind of the fruit and underside of the pulp.

INSECT PEST OF POMEGRANATE

POMEGRANATE BUTTERFLY

Virachola isocrates (Fabricius) is economically the most important of all the butterflies, perhaps the only one that is constantly and regularly injurious. It is a polyhphagous pest having a very wide range of host plant, including, aonla, apple, ber, citrus, guava, litchi, loquat, mulberry, peach, pear, plum, pomegranate, sapota and tamarind. It is widely distributed all over India and is found wherever pomegranates are grown.

The female lays eggs single on calyx of flowers or small fruits. On hatching, the caterpillars bore inside the developing fruits and are usually found feeding on pulp and seeds just below the rind. As many as eight caterpillars may be found in single fruit. Subsequently, the infested fruits are also attacked by bacteria and fungi causing the fruits to rot. The conspicuous symptoms of damage are offensive smell and excreta of the caterpillars coming out of entry holes, the excreta is found struck around the holes. Sometimes the holes may also be seen plugged with the anal end of a caterpillar. The affected fruits ultimately fall down and are of no use; even if the fruits be picked before falling down; these fruits have no market value.

Eggs are shiny white in colour and oval in shape. Full grown caterpillars are stout, 17 to 20 mm long, dark brown in colour with short hair and whitish patches all over the body. Just before pupation, the caterpillars come out of the fruits and tie the stalk with main branch of the tree with fine silken strands, to ensure that the fruit does not fall down, they re-enter the fruit and pupate there in. Occasionally, the caterpillars may pupate outside also, attaching themselves to the stalk of the fruits. Adult butterflies are medium-sized, glossy-bluish-violet (males) to brownish violet (females) in colour; females have conspicuous orange patch on the fore wings. Wing length is 40 to 50 mm.

Incubation, larval and pupil periods range from 7 to 10, 18 to 47 and 7 to 34 days, respectively without overlapping generations in a year. The pest breeds throughout the year.

To reduce the incidence of this pest, remove and destroy all the affected fruits. The only other effective though expensive method in bagging of fruits and may be practiced, if the number of fruits in trees is limited. It is suggested to spray with 0.03% phosphamidon just when the fruit formation starts.

Pomegranate fruits are also damaged by pomegranate borer, Deuodorix isocrates (Moore), castor capsule borer,

Dichocrocis punctiferalis Guenee, pyralid bore, *Euzophera punicaella* Moore and fruit sucking moths; *Othreis fullonia* clerk and *O. materna* (Linnaeus). Besides, Ayyar (1924) reported two pentatomid bugs, *Jurtina indica* Dallas and *Halyomorpha picus* (Fabricius) puncturing the tender and ripe fruits. These cause only minor damage to pomegranate fruits. Except *D. epijarbas*, others are polyphagous pests and cause major damage to other economic crops. *D. epijarbas* has been occasionally causing serious damage, specially in Himachal Pradesh and Uttar Pradesh.

BARK EATING CATERPILLARS

Indarbela tetraonis and *I. quadrinotata* have been recorded boring the bark of pomegranate trees and feeding inside. *I. tetraonis* is comparatively more common and more harmful to pomegranate trees that *I. quadrinotata*. Older trees and the trees in orchards that are not well maintained are more prone to these pests. Usually there is only one caterpillar in each hole but there may be 10 to 12 holes in a badly infested tree. Such trees will bear no fruits.

Besides *Indarbela* spp., the red borer, *Zeuzera coffeae* Nietner has also been reported boring the stems and trunks of pomegranate trees. This is a polyphagous pest, coffee and tea being its preferred hosts. It is a minor pest of citrus, custard apple, guava, lithchi, loquat, pomegranate, etc. A female lays 500 to 1000 eggs in 1 to 2 weeks. Eggs are laid in strings on the bark, branches or stems. On hatching, the larvae bores, usually at the joints between leaf-stalk or twig the and main stem. And tunnel straight downwards. The larvae also cuts circular holes at various places through which they eject the frass etc. Pupation takes place inside the tunnels. Eggs are oval (1.0 x 0.6mm) in shape and reddish-yellow in colour. Caterpillars are pinkishwhite with dark brown spots and are about 40 mm long when full grown. Pupae are chestnut-brown in colour and 22 to 28 mm long. Moths are white with pairs of small black dots on thorax; numerous small black spots and streaks on fore wings and few black spots on posterior edges of hind wings. Wing expanse is 35 to 45 mm. Eggs hatch in about 10 days. Larval development takes 60 to 120 days and pupal period lasts for 3 weeks to one month. Total lifecycle lasts for 4 to 5 months in South India and at low elevations, and extends to more than a year at high elevations in North India.

Keeping the orchards, clean and avoiding over-crowding of trees helps in minimizing the attack by these borers. In case of infestation, clean the affected portions by removing all webs, etc, and insert in each hole, swab of cotton-wool soaked in carbon bisulphide, petrol or even kerosene and seal and holes with mud.

STEM BORING BEETLES

Celosterna spinator Febricius and *Olenecamptus bilobus* Fabricius have been reported boring the stems and trunk of pomegranate trees. Both are polyphagous pests causing minor damage. *O. bilobus* is widely distributed in the oriental region. It is primarily a pest of Ficus spp. It prefers breeding in dead wood but also attacks the living branches. The grubs bore inside the trunk and feed on sapwood. Adult beetles are active by the day and feed by gnawing the green bark of shoots. The life-cycle is annual with an extended emergence period from May to November.

C. spinator is major pest of babul, *Acacia arabica*. Adult beetles have pale yellowish-brown body with light grey elytra and are 30 to 35 mm long. These are nocturnal in habit and appear with onset of monsoon; emergency is accelerated by continuous heavy rains. Egg period is 12 to 15 days, grub 9 to 10 months and pupal 16 to 18 days. There is only one generation in a year and longevity of beetles is 45 to 60 days.

Generally no control measures are required. Nevertheless, the measures suggested against bark eating caterpillars will prove effective in controlling these beetles as well.

SAP SUCKING INSECTS

Drosicha mangiferae (Green), *Hemiaspidoproctus cinerea* (Green), *Icerya purchasi* Maskell, *Planococcus lilacinus* (Cockerall), *Andaspis hawaiiensis* (Maskell,) *Aonidiella orientalis* (Newsstead), *Duplaspidiotus tesseratus* (de Charmoy,) *Hemiberlesia punicae* (Signoret), *Lindingaspis greeni* (Brain and Kelly), *L. rossi* (Maskell), *Parlatoria oleae* (Colvee), *Pinnaspis theae* (Maskell) are some of the mealy bugs and scale insects reported from various part of India on pomegranate leaves. Most of these are polyphagous, having a wide range of host plants. Due to damage by these insects, the trees are devitalized, resulting in shedding in buds and flowers and smaller sized fruits. To prevent the attack from spreading, prune and destroy the affected parts in the initial stage of attack. If the infestation becomes severe, spray with 0.04% diazinon or monocrotophos.

Pomegranate aphid *Aphis punicae* and pomegranate whiteflies Siphoninus finitimus Silvestri and *S. Phillyreae* (Haliday) are also minor pests recorded occasionally. They feed by sucking the cell sap from leaves and tender twigs. The affected parts get discoloured and disfigured. These insects also secrete copious amount of honeydew on which sooty mould develops, hindering the photosynthetic activity of the plant.

Thrips, *Retithrips syriacus* (Mayet) and *Rhipiphorothrips creuentatus* Hood have been recorded feeding on leaves, while *Anaphothrips oligochaetus* Karyny. *Ramaswamiahiella subnudula* Karny and *Scirtothrips dorsalis* Hood infest the flowers. These are all minor pests of pomegranate. Nymphs and adults lacerate the leaves, flower stalks, petals and sepals and rasp the sap that oozes out of these wounds. As a result, the leaf tips curl and dry away while flowers are shed and ultimately the fruiting capacity of the tree is adversely affected.

To control the aphid, whiteflies and thrips, spray with 0.03% dimethoate, oxydemeton methyl or phosphamidon.

Two spraying at an interval of 10 to 15 days will be sufficient to keep complete check of these pests.

LEAF EATING CATERPILLARS

Leaves of pomegranate trees are often attacked by one or the other lepidopterous larvae. Those of regular occurance include, castor semilooper. Hairy caterpillars, slug caterpillars and bag worms.

Castor semilooper, *Achaea janata* (Jannaeus) is an important pest of castor. It has also been reported damaging ber, citrus, grapevine, guava, pomegranate, etc. The pest has countrywide distribution. Eggs are laid singly on tender leaves usually on ventral surface, one to six eggs may be found on a leaf. A female lays on an average 400 eggs. Freshly hatched caterpillars congregate on the leaves of various weeds and few economic crops, including pomegranate, ber and castor, and feed on the chlorophyll. Later, they segregate and feed voraciously devouring the entire leaf lamina. In case of severe infestation, the young tree may be completely defoliated. The moths are also destructive. They suck the juice of various fruits, like citrus, grapes, guava, mango etc. Eggs are hemispherical bluish-green and ridged with 40 to 45 striae. Caterpillars are semiloopers about 5 mm when freshly hatched become 55 to 65 mm when fully grown.

The caterpillars show conspicuous colour variations, some are grey with red or brown lateral stripes while others are bluish-grey specked with blue-black and having yellowish lateral stripes. Pupation takes place in the soil 40 to 50 mm deep or even among the fallen, leaves or in folds of leaves. Freshly formed pupae are glistening dark green, later becoming brown. Moths are stout, pale reddish-brown with wavy lines on fore wings and black hind wings having a medial white bank and three large white spots on the outer margin. Anal and apical margins of the wings are fringed with hair. Wing expanse is 50 to 65 mm.

Incubation period is 3 to 5 days while the larval and pupal

periods last for 9 to 23 days and 7 to 26 days respectively. Preoviposition period of moths extends from 6 to 21 days and life-cycle occupies 3 to 5 weeks during active period of the pest. There are five to six overlapping generations in a year and wintering takes place in the pupal stage.

To check the damage caused by these semiloopers, collect and destroy the caterpillars mechanically. In case of severe infestation, dust 50% Malathion or spray 0.05% dichlorvos or endosulfan. This will kill other leaf eating caterpillars, if any.

Ber hairy caterpillar, *Euproctis flava* (Bremer), plum hairy caterpillar, *E. fraterna* (Moore) and castor hairy caterpillar, *E. lunata* Walker are the species reportedly damage pomegranate trees. These are highly polyphagous pests, found all over the Indian sub-continent. Eggs are laid in clusters on ventral surface of leaves and covered with hair. On hatching, the caterpillars feed gregariously on eqidermis of leaves; later they segregrate and feed voraciously, defoliating the entire trees. Pupation takes place in hairy cocoons on leaves or on branches. Moths of *E. flava*, have head and thorax bright orange-yellow and abdomen pale. Fore wings are orange yellow in colour with orange spots at cells and three sub-apical black spots; hind wings and paler than forewings. Wing expanse is 30 mm to 40 mm. Moths of *E. fraterna* are more or less similar to those of *E. flava*, except that these are smaller in size, wing expanse being 24 mm to 34 mm. *E. lunata* moths are pale in colour with white fore wings having a large black lunule; wing expanse is 34 mm to 38 mm.

Biology in these species is more or less the same. The eggs hatch in 5 to 10 days, larval period lasts for 4 to 5 weeks and pupal stage occupies 10 to 12 days—thus the entire life-cycle is completed in 45 to 57 days.

Porthesia scintillans is another hairy caterpillar widely distributed in the Indian sub-continent. This is also a polyphagous pest, damaging, apple, mango, pomegranate etc.

Caterpillars are dark brown with a series of crimson lateral tubercles on a yellow line bearing tufts of grey hair. Moths are yellowish with reddish line and spots on the edges. Fore wings are vinous-brown irrorated with dark scales and hind wings are fuscous-brown with yellow margin; wing expanse is 20 to 26 mm and 32 to 38 mm in case of males and females, respectively.

Egg, larval and pupal periods last for 6 to 10, 30 to 40 and 8 to 12 days respectively.

Arcyophora dentula Hampson has been reported as a minor pest. The caterpillars burrow holes in the leaves and ultimately defoliate the tree. Maximum damage is caused during September-October.

Caterpillars of *Creatonotus gangis* (Linnaeus) have been reported feeding on leaves during January to March in Andhra Pradesh.

Egg, larval and pupal periods last for 3 to 4,30 to 33 and 10 to 13 days respectively.

Dusting with 5% Malathion is effective for checking these hairy caterpillars. It is desirable to carry out the campaign on a co-operative basis and during early stage of infestation when the caterpillars are feeding gregariously; at this stage, even 5% Malathion dust will increase the mortality rate of these caterpillars.

Bag worms are widely distrubuted all over India. The species recorded feeding on pomegranate leaves, include, *Clania creameri* Westwood and *Acantho psyche* . The caterpillars construct a case over their body and live within, nibbling leaf lamina. Females are apterous and devoid of antennae, mouth parts and legs. Males are winged and have bipectinate antennae and short mouth parts.

BEATLES AND WEEVEILS

Anomala dimidiate Hope, *Hoplasoma sexmaculata* Hope, *Mimastra cyanure* Hope, *Myllocerus laetivirens* Marshall and *M. undecimpustulatus maculosus* Desbrocher have been occasionally reported feeding on pomegranate foliage. These are polyphagous pests, destructive mostly to temperate fruits, trees and are of minor importance in case of pomegranate. The grubs live in the soil and feed on the roots and other organic matter. It is only the adults that come out of the soil at night and feed on foliage. Generally no control measures are required against these pest, but if and when necessary, dust with 5% Malathion.

FRUITS FLIES

Mango fruit fly, *Dacus dorsalis* (Hendel) and peach fruit fly, *D. zonatus* (Saunders) have been reported attacking pomegranate fruits as well, though the loss caused is of minor importance. These flies are cosmopolitan in distribution and feed on a vast variety of fruits and vegetables. Both the species are found attacking apple, bael, ber, citrus, fig, guava, mango, peach, pomegranate, sapota, etc. In additon, *D. dorsalis* has also been recorded on apricot, banana, jact-fruit, laquat, pear, persimmon, plum, etc., while *D. zonatus* has been recorded damaging custard apple.

Eggs are laid just below the epidermis of ripening fruits. The affected fruits show dark brown punctures through which juice oozes out. On hatching, the small white grubs feed inside the fruits, as a result of which the fruits rot and fall down.

To prevent population build-up and carry-over of these flies, collect and destroy promptly all the fallen and affected fruits. To avoid the infestation, harvest the fruits before they ripen as the flics are not able to puncture the hard rind of unripe fruits.

TERMITES

Odontotermes obesus (Rambur) has been reported attacking pomegranate trees in Rajasthan and Andhra Pradesh. This subterrranean enemy is highly polyphagous and feeds on a large number of economic and wild crops. Its attack is more pronounced in sandy soils than in clay or black soils. The termites cannot thrive under conditions of bad aeration and poor drainage

Soil application with 5% aldrin, chlordane or heptachlor dust is quite effective in warding off the attack of this pest and should be regularly practiced at least once in a year in termite infested orchards.

SCAVENGERS

Peach scavenger moth, *Anatrachyntis simplex* willingham has been found attacking pomegranate fruits specially over-ripe and malformed ones or those that have been attacked by fungi or bacteria but are still hanging on trees. The caterpillars web together a few seeds with silken stands and feed within.

Scavenger beatles (calvicorn) feed on fermenting or decaying vegetable matter, particularly those with exuding sap, souring fruits and withering flowers, decomposing bark and sapwood. The species recorded on pomegranate are, Amphicrosus species and *Carpophilus dimidiatus* Linnaeus. These polyphagous beatles feed on rotten fruits that have been already damaged by the caterpillars of pomegranate butterfly or fruit flies and have fallen down. *C. dimidiatus* has been reported feeding on damaged fruits of guava, mango, peach, pear, plum, etc. To check the population of these scavengers collect all rotten and infested fruits and destroy the same promptly.

ORGANIC FARMING

Most countries have traditionally utilized various kinds of organic materials to maintain or improve the tilt, fertility and productivity of their agricultural soils. However, several decades ago organic recycling practices in some countries were largely replaced with chemical fertilizers which were applied to high-yielding cereal crops that responded best to a high level of fertility and adequate moisture, including irrigation. Soil tillage was also intensified to improve weed control and seedbed conditions. Consequently, the importance of organic matter to crop production received less emphasis and its proper use in soil management was sometimes neglected or even forgotten.

Soil Organic Matter or Humus

Farmers since ancient times have recognized significant benefits of soil organic matter to crop productivity. These benefits have been the subject of controversy for centuries and some are still debated today. The following list includes many of the recognized benefits of soil organic matter.

1. It serves as a slow-release source of N, P, and S for plant nutrition and microbial growth.
2. It possesses considerable water-holding capacity, and thereby helps to maintain the water regime of the soil.
3. It acts as a buffer against changes in pH of the soil.
4. Its dark colour contributes to absorption of energy from the sun and heating of the soil.
5. It acts as a *cement* for holding clay and silt particles together, thus contributing to the crumb structure of the soil and to resistance against soil erosion.
6. It binds micro-nutrient metal ions in the soil that otherwise might be leached out of surface soils.
7. Organic constituents in the humic substances may act as plant-growth stimulants.

TYPES OF BULKY ORGANIC MANURES

Farm yard manure (FYM), farm compost, town compost, night soil, sludge and green manures are bulky in nature and supply large quantities of organic matter but small quantities of plant nutrients in comparison to the inorganic fertilizers.

Farmyard Manure

Good-quality farmyard manure is perhaps the most valuable organic matter applied to a soil. It is the most commonly used organic manure in most countries of the world. It consists of a decomposed mixture of cattle dung, the bedding used in the stable and any remnants of straw and plant stalks are fed to cattle. It is one of the most important agricultural by-products. Unfortunately more than 50 per cent of the cattle dung produced in India is burnt as agriculture.

Table - Average nutrient content of bulky manures

Manure	Percentage content		
	Nitrogen (N)	Phosphoric acid P_2O_5)	Potash (K_2O)
Animal refuse			
Cattle dung, fresh	0.3-0.4	0.1-0.2	0.1-0.3
Horse dung, fresh	0.4-0.5	0.3-0.4	0.3-1.0
Sheep dung, fresh	0.5-0.7	0.4-0.6	0.3-1.0
Nightsoil, fresh	1.0-1.6	0.8-1.2	0.2-0.6
Poultry manure, fresh	1.0-1.8	1.4-1.8	0.8-0.9
Raw sewage, fresh	2.0-3.0	-	-
Sewage sludge, dry	2.0-3.5	1.0-5.0	0.2-0.5
Sewage sludge, activated dry	4.0-7.0	2.1-4.2	0.5-1.0
Cattle urine	0.9-1.2	Tr.	0.5-1.0
Horse urine	1.2-1.5	Tr.	1.3-1.5
Human urine	0.6-1.0	0.1-0.2	0.2-0.3
Sheep urine	1.5-1.7	Tr.	1.8-2.0
Wood ashes			
Ash, coal	0.73	0.45	0.53
Ash, household	0.5-1.9	1.6-4.2	2.3-12.0
Ash, wood	0.1-0.2	0.8-5.9	1.5-36

Farm, factory and habitation wastes

Rural compost, dry	0.5-1.0	0.4-0.8	0.8-1.2
Urban compost, dry	0.7-2.0	0.9-3.0	1.0-2.0
Farmyard manure, dry	0.4-1.5	0.3-0.9	0.2-1.9
Filter-press cake	1.0-1.5	4.0-5.0	2.0-7.0
Plant residues			
Rice hulls	0.3-0.5	0.2-0.5	0.3-0.5
Groundnut husks	1.6-1.8	0.3-0.5	1.1-1.7
Straw and stalks			
Pearl millet	0.65	0.75	2.50
Banana, dry	0.61	0.12	1.00
Cotton	0.44	0.10	0.66
Sorghum	0.40	0.23	2.17
Maize	0.42	1.57	1.65
Paddy	0.36	0.08	0.71
Tobacco	1.12	0.84	0.80
Pigeon pea	1.10	0.58	1.28
Wheat	0.53	0.10	
Sugarcane	0.35	0.10	0.60
Tobacco dust	1.10	0.31	0.93
Tree leaves, dry			
Calotropis gigatean	0.35	0.12	0.36
Careua arbprea	1.67	0.40	2.20
Cassoa auriculata	0.98	0.12	0.67
Dillenia Pentagyna	1.34	0.50	3.20
Madhuca indica	1.66	0.50	2.00

Pengamia pumata	3.69	2.41	2.42
Pterocarpus marsuptum	1.97	0.40	2.90
Terimalia chebula	1.46	0.35	1.35
Terminalia paniculata	1.70	0.40	1.60
Terminalia tomentosa	1.39	0.40	1.80
Xylia dolbriformis	1.37	0.30	1.61
Green manures, fresh			
Cow pea (Vignaa catjang)	0.71	0.15	0.58
Sesbania aculeate	0.62	-	-
Cluster-bean(Cyamopsis tetragonoloba)	0.34	-	-
Horse-gram(Dolichos biflorus)	0.33	-	-
Moth bean (Vigna aconitifolia)	0.80	-	-
Green gram (Vigna radiate)	0.72	0.18	0.53
Sunnhemp (Crotalaria juncea)	0.75	0.12	0.51
Black gram (Vigna mungo)	0.85	0.18	0.53

Composted Manure

Another method of agumenting the supplies of organic matter is the preparation of compost from farmhouse and cattle-shed wastes of all types. Composting has been advocated and adopted extensively during the past 25 years. composting is the process of reducing vegetable and animal refuse (rural or urban) to a quickly utilizable condition for improving and maintaining soil fertility. Studies conducted in India and elsewhere have shown that good organic manure similar in appearnace and of fertilizing value to cattle manure can be produced from waste materials of various kinds, such as cereal straws, cotton stalks, groundnut husk, farm weeds and grasses, leaves, leaf-mould, house refuse, wood ash, litter, urine-soaked earth from cattle-sheds and other similar substances.

These material are rich in cellulose and other readily decomposable carbohydrates and have a carbon-nitrogen ration of 40 or more than 1. The direct application of such under-composed, low-nitrogen organic matter as manure brings about a temporary deficiency of mineral nutrients (specially nitrates and ammonium compounds) in the soil by stimulating the growth of micro-organisms, which in turn, compete with crop plants for available nitrogen, phosphorus and other elements. Hence, before using them as manure, it is necessary to compost or partially decompose them. This process lowers the carbon-nitrogen ratio to about 10 to 12 is to 1.

Urban Compost

In recent years large-scale composting of town refuse and night soil in properly constructed trenches, away from human habitation, has been taken up successfully by the municipal bodies of many large and small towns. Trenches, 1 to 1.2 m wide, 75 cm deep and of convenient length, are filled with successive layers of night-soil, town refuse and earth, in respective order. The compost then gets ready in about three months.

Sewage and Sludge

Domestic and industrial wastes (sewage and sludge) contain large quantities of plant nutrients and are used for growing of crops near many towns. In many places, the undiluted silage has been found to be too strong for healthy plant growth and if it contains readily oxidizable organic matter, its use actually reduces nitrates present in the soil. The disadvantages are still greater if sewage is used on land without preliminary treatment. The soil quickly becomes sewage sick owing to the mechanical clogging by colloidal matter in the sewage and the development of the soil but also produce alkalinity. Bacterial contamination makes the eating of raw vegetables grown on untreated sewage or sewage sludge a real danger to health.

Municipal and sewage wastes also form an important component of organic farming. The total and utilizable nutrient potential from garbage and sewage sludge works out to be substantial. The total municipal refuse is about 12 million tonnes/annum containing about 0.5% N, 0.3% K, whereas sewage sludge amounting to 4 million tonne/annum contains about 3% N, 2% P and 0.3% K.

Besides, the necessity of sewage farming from nutrient utilization point of view, is also an effective method to avoid pollution.

Night soil

Very few towns in India are equipped with proper sewage system. The sanitary disposal of night soil with an effective control of foul smell and fly nuisance is therefore, a serious problem all over the country. Since human excreta is a potential source of soil imrpovement, public health authorities in several countries make the necessary arrangement for its conservation and conversion into a form in which it can be safely used as a manure. The dehydration of night soil, as such, or after admixture with absorbing materials, e.g. soil ash, charcoal, and sawdust, produces a poudrette that can be easily used as a manure. The mixing of night soil with an equal volume of ash and 10 per cent powdered charcoal produces an odourless material, containing 1.32 per cent nitrogen, 2.8 per cent phosphoric acid 4.1 per cent potash and 24.2 per cent lime. The addition of 40 to 50 per cent of sawdust of the night-soil yields straight away a dry, acidic poudrette which may contain 2 or 3 per cent nitrogen.

Green manuring

Green manureing can be defined as a practice of ploughing or converting the soil recompensed green plant tissues for the purpose of improving physical structure as well as fertility of the soil. From time immemorial the convertin to a green crop for improvement of the condition of the soi has been a popular farming practice. Green manuring, wherever feasible, is the principal supplementary means of adding organic matter to the soil. It consists in the growing of a quick growing crop and phoughing it under to incorporate it into the soil. The green manure crop supplies organic matter as well as additional nitrogen, particularly if it is a legume crop, which has the ability to fix nitrogen from the air with the help of its root-nodule bacteria. The green manure crops also exercise a protective action against erosion and leaching.

The adoption of green manuring depends upon the agro-climmatic conditions. Broadly the following two types of green manuring can be thought of:

1. **Green manuring in situ:** In this system, green manure crops are grown and buried in the same field which is to be green-manured, either as a pure crop or as an inter-crop with the main crop. In India, various methods of growing green manure crop in situ are followed to suit local conditions. For the proper decomposition of the green manure, it is necessary that the green material should be succulent and there should be adequate moisture in the soil. Plants at the flowering stage, contain the greatest bulk of succulent organic matter with a low carbon/nitrogen ration. The incorporation of the green-manure crop into the soil at that stage allows a quick liberation of nitrogen in the available form.

2. **Green leaf manuring:** Green-leaf manuring refers to purning into the soil green leaves and tender green twigs collected from shrubs and trees grown on bunds, waste lands and nearby forest areas. The common shrubs and trees used are Glyricidia (Glyricidia maculate), Sesbania speciosa, Karanj (Pongamia pinnata), etc.,

Green manuring in situ is followed in northern India while green leaf manuring is common in eastern and central India.

Concentrated Organic Manures

Concentrated organic manures are those materials that are organic in nature and contain higher percentages of essential plant nutrient such as nitrogen, phosphorus and potash, as compared to bulky organic manures. These concentrated manures are made from raw materials of animal or plant origin. The concentrated organic manures commonly used are oil cakes, book-meal, fish meal, meat and horn and hoof meal.

Oil cakes

These can be grouped into two classes, namely, (i) edible oil cakes—suitable for feeding the cattle and (ii) non-edible oil cakes—not suitable for feeding the cattle. Edible oil cakes, are normally fed to cattle as concentrates. However, some of the edible oil cakes are also applied by the farmers to the soil. The quantity of nitrogen varies with the type of oil cake. It is from 2.5 per cent in mahua cake to 7.9 per cent in decorticated safflower cake. In addition to nitrogen, all oil cakes are the quick-acting organic manures. Though they are insoluble in water, their nitrogen becomes quickly available to the plants in about 7 to 10 days after application. The mahua cake, however is an exception, as its nitrogen content does not become available till about two months after application. As such mahua cake should be applied about two months before sowing, provided the soil is moist. This cake is suitable for application to fruit plants and plantation crops.

How to use oil cakes? These should be well-powdered before application, so that there can be spread evenly and are easily decomposed by micro-organisms. They can be applied a few days prior to sowing or at sowing time. Oil cakes—especially groundnut cake, are also applied extensively in the form of a top-dressing to Sugarcane. Depending on crop, oil cakes are applied broadcast, drilled or placed while earthing up near and root zone.

Cattle, Pig, and Poultry Manures

In general the higher the dry matter, the higher the nutrient content. For example, in Table the higher nutrient content of the poultry manure is due mainly to the higher dry matter content. There is usually about the same level of nitrogen and potassium in cattle slurry, and phosphorus content is usually about 25% of the nitrogen and potassium level. Pig manure usually has about the same level of nitrogen and potassium in cattle slurry and phosphorus content is usually about 25% of the nitrogen and potassium level.

Pig manure usually has about the same level of nitrogen as does cattle manure, although pig manure is usually higher in phosphorus and lower in potassium than is cattle slurry. The difference in the phosphorus to potassium ratio between cattle and pig manure is principally the reflection of the diet of these animals. The relatively low potassium and high phosphorus in pig slurry reflects mainly a cereal diet and the high potassium in cattle manure reflects the high potassium content of her bage which often makes up a high proportion of the ruminant diet.

Meat meal

An adult animal can provide 35 to 45 kg of meat after slaughter or death. At present meat meal is manufactured in India on a small scale. The process of manufacturing meatment is very simple. First, the bones and the meat are cooked or digested in a special container for two to three hours. The bones are then separated from the meat. This meat is dried on a sand bath till it is brittle and them it is powdered. The drying can be done in double-jacket trays worked by steam and the material dried over the steam for an hour. Mutton squeezer can also be used for removing water content of the meat and then dried. Meatmeal is a quick-acting manure and is effective for all crops on all soil types. Its application is similar to oil cakes. Meatmeal contain about 10.5 per cent N and 2.5 per cent P_2O_5.

Blood meal

An adult cattle gives about 14.0 kg. and a goat or sheep about 1.40 kg of blood. Slaughter houses should be provided with pucca or concrete floor with a central drain leading into a blood storage tank. The blood is first treated with commercial copper sulphate@ 125 gm per 100 kg of blood. It is then evaporated to dryness on a sand bath. Next, it is spread on a concrete floor covered by a net, and allowed to dry in under the sunlight. When completely dried, it is powdered, bagged and marketed as blood meal. Blood meal is a quick-acting manure and is effective for all crops on all soil types. It should be applied like oil cakes. Blood meal contains 10 to 12 per cent of N and 1 to 2 per cent of P_2O_5.

Fish meal

An edible fish carcasses are used to prepare fish meal. The fish is dried, crushed or powered and filled in bags. Fish manure is available either as a dried fish or as fish meal or powder. The manorial constituents present in it vary with the type of fish. Fish meal is quick-acting organ manure and is suitable for application to all crops on all soils. It should preferably by powdered before use. Fish meal contains 4 to 10 per cent P_2O_5 and 0.3 to 1.5 per cent K_2O.

Horn-and-hoofmeal

A healthy animal gives about 3 to 4 kg of horn and hoof. These materials are cooked dried and powdered before bagging. This meal contains about 13 per cent N.

Macro and Micro-nutrient Availability

Mineralization of plant nuitrients and their availability of plants is one of the major consequence of organic recycling in soil. The quantity released depends mainly on the content of a particular nutrient in organic residues undergoing decomposition. In general there is immobilizatioin of nitrogen due to decomposition of cereal straw. However, with FYM/ compost application no such effect is observed. The addition of FYM and cereal residues results in improvement of total soil nitrogen. The effect of organic matter in reducing the intensity of phosphate fixation by the soil sequences and maintenance of soil fertility by use of organic manures along with super phosphate this has also been established. FYM remains intermediate in building up available P status of soil. In sub-humid later tic soils, P use efficiency under rice-rice sequence was increased tremendously with FYM application. Information needs to be generated on what amount of fertilizer P could be curtailed for organic manures under different cropping sequences. Response of crops to organic manureing depends on degree of decomposition of organic residues, C:N ratio. Time of application and soil characteristics.

Continuous application of organic manures improves the availability of Zn in soil and may not be sufficient to meet the requirements of the immediate crops. Contrary to macro-nutrients, the range between deficiency and toxicity limits of micro-nutrients is quite narrow for most of the crops. Inclusion of micro-nutrients in fertilizer schedule should, therefore, be advocated after careful appraisal through soil and plant tests.

Availability of organic waste in India

There are various estimates of agricultural waste availability. These may be based on crop residues, net availability of which on remote sensing techniques is used in developed countries.

Estimates of agricultural waste availability suggest that the average value for crop wastes is 350 mt and that of animal wastes is 650 mt. Hence, around 1000 mt at agricultural wastes are available in the country.

Crop Waste: The quantity of wastes produced daily by an animal is a function of the type of animal, its size, its feed, and the temperature and humidity of the environment. Quantities quoted in the literature often refer to total weight or volume or fresh waste produced, including variable quantities of water. Intensive livestock rearing tends towards standardization of diet and type of animal. **Cattle:** The slurry production rate most often quoted for cattle is the region of 40 to 45 kg per day per livestock unit. A livestock unit (L.U) is equivalent to a 540 kg dairy cow. 500 kg beef cattle produces 40 kg. of undiluted slurry per animal per day. It has been estimated that about 3 tonnes of dung stead manure or 6 tonnes of farm yard manure would be produced by one livestock unit during a 140-day winter period.

Table: Agricultural waste availability in India.

Agricultural waste	Quantity (mt.)
Crop waste (including industrial waste)	301.6
Animal waste (including fisheries is and marine waste)	944.4
Crop residues (including industrial waste)	108.0
Crop waste (cereals only)	289.7
Crop residues	407.0
Animal wastes	2018.0
Crop residues (including industrial waste)	287.3
Crop waste residues	287.3
Animal waste	633.0

Pigs: The most often quoted slurry production for meal-fed pigs is 4.5 kg per pig per day on the average over the flattening period. At an average weight of 25 kg. per pig. 4.35 kg slurry composed of 1.73 kg feceas and 2.62 kg urine was produced. Pigs at 25 kg. mean weight produced 4.5 kg slurry per day with 10% dry matter. One piglet produces about 14.3 kg slurry per day.

Poultry: Daily fresh poultry manure represents about 5% of bird live weight. A 2 kg mature bird would produce about

0.1 kg manure daily with 25% dry matter. Broiler droppings are mixed with litter and normally contain about 75% dry matter when removed from the house.

Silage Effluent: The quantity of silage effluent from ensiled grass and other materials is usually related to the dry matter of the ensiled materials; the lower the dry matter the greater the quantity of effluent. A tonne of ensiled material with 10 to 20% dry matter will produce 260 to 450 kg of effluent. At 20% dry matter, approximately 100 kg effluent per tonne is produced.

Growing Mushrooms: Spent compost from growing mushrooms, alongwith other composts are spread on the land and and have a significant fertilizer value. The quantities are small by comparison with other agricultural wastes; however, they can be important in localized areas..

Utilization of Waste

- Chemical constituents like cellulose, hemi-cellulose and lignin lead to establish the potential agricultural wastes and food, fee, fodder, fiber and chemical products of economic importance. The calorific value (CV) indicates the fuel utility of agricultural waste. The CV for various agricultural waste materials varies between 14 to 19 MJ/kg. Higher C content favours calorific value of the waste material. The volatile water in plant residues may be around 80 per cent and fixed carbon is 20 per cent. Volatile matter in plant residues contribute towards oil and tar content. The characteristics of common agricultgural wastes available in India:
- The characteristics of the agricultural waste materials make them important in respect of their use and economic value. Agricultural wastes can be put to use because of the following reasons:
- Farm development (improvement of soil health, soil fertility, soil-physical conditions and plant protection),

It is the source of energy and power,

- It acts as animal feed and fodder,

1. It is the human food.

Table: Elemental composition of common agricultural wastes

Agricul-tural waste (biomass	Nutrient Content (%)								
	C	H	N	Na	K	P	Ca	Mg	SiO_2
Pigeon pea stalks	53.30	4.70	0.60	0.05	0.57	0.08	0.11	0.41	0.68
Bagasse	48.20	6.10	0.20	0.06	0.51	0.04	0.14	0.36	1.30
Totton sticks	51.00	4.90	1.00	0.09	0.61	0.08	0.12	0.43	1.33
Groundnut Shell	41.10	4.90	1.60	0.05	1.20	0.12	0.10	0.40	2.52
Maize cobs	46.10	4.90	0.60	0.03	0.54	0.07	0.08	0.28	0.90
Maize stalks	41.10	4.20	0.60	0.04	0.42	0.05	0.09	0.45	2.00
Rice husk	37.80	5.00	0.30	0.02	0.30	0.03	0.08	0.17	15.60
Rice straw	36.80	5.00	0.59	0.09	2.50	0.06	0.10	0.53	16.77
Wheat straw	43.80	5.40	0.40	0.06	0.78	0.04	0.10	0.35	7.08

BIO-FERTILIZERS/MICROBIAL INOCULANT FOR PRODUCTION OF HORTICULTURAL CROPS

Some soil micro-organisms play an important role in improving soil fertility and crop productivity due to their capability to fix atmospheric nitrogen, solubilise insoluble phosphate and decompose farm wastes resulting in the release of plant nutrient. The extent of benefit from these micro-organisms depends upon their number and efficiency which, however, is governed by a large number of soil and environmental factors. When the number and activity of special micro-organism called microbial inoculants or bio-fertilizer is used, it hastens biological activity to improve availability of plant nutrient. A number of products are now available that are generally referred to as soil and plant additives, of non-traditional nature. These products include (i) microbial fertilizers and soil inoculants which are purported to contain unique and beneficial strains of soil micro-organisms, (ii) microbial activators that supposedly contain special chemical formulations for increasing the numbers and activity of beneficial micro-organisms in soil, (iii) soil conditioners that claim to create favourable soil physical and chemical conditions which result in increased growth and yield of crops, and (iv) vermin-compost which helps in improving soil health and fertility.

Nitrogen fixing organisms can be provided to be farmers in the name of microbial inoculants otherwise termed as Bio-fertilizers. The bio-fertilizers containing biological nitrogen fixing organisms are of utmost importance in agriculture in view of the following advantages:

1. They help in the establishment and growth of crop plants and trees.
2. They enhance bio-mass production and grain yields by 10-20%.
3. They are useful in sustainable agriculture.
4. They are suitable in organic farming.
5. They play an important role in agro-forestry/silvipastoral systems.

After the advent of the acetylene test for assessing nitrogen fixing potential, rapid progress has been made in identifying nitrogen fixing organisms. Out of a large number of micro-organisms possessing the property of nitrogen fixation, only a few such as Rhizobium, Azotobacter, Azosirillum, BGA, Azolla, etc. have been commercially exploited. The reaction in biological nitrogen fixation is essential, the same as in production of chemical fertilizers (Haber-Bosch process) i.e.s the catalytic reduction of dinitrogen. (N_2) to ammonia (NH_3).

Types of Bio-fertilisers

Rhizobium Inoculants

The most widely used bioferliser is Rhizobium which colonizes the roots of specific legumes to form tumors-like growths called root nodules. It is these nodules that act as factories of ammonia production. Rhizobia have the ability to fix atmospheric nitrogen in symbiotic association with legumes and certain non-legumes like. Parasponia Rhizobia colonize the roots of specific legume, enter through root hair's multiply and form tumor-like growths called root nodule. A mature nodule consists of a central bacteroid zone surrounded by several layers of cortical cells. The process of N-fixation is wholly dependent on the activity of the enzyme nitrogenous which is located within the bacteriods. The volume and number of asteroids have a direct positive relationship with the amount of nitrogen fixed.

The Rhizobium-legume association can fix up to 100-300 kg N/ha in one crop season and in certain situations can leave behind substantial nitrogen for the following crop. The range of nitrogen fixed per hectare per year by different legumes is 100-320 kg for fiber bean, 90-100 kg for lentil, 150-200 kg for lupins, 50-60 kg for groundnut, 60-80 kg for soybean, 50-55 kg for mungbean and 100-400 kg for pasture legumes.

Azotobacter Inoculants

Azotobacter is one of the most important non-symbiotic N-fixing micro-organism and considered to be very important for fixation of nitrogen in non-leguminous plants. Based on morphological and physiological features, the genus Azotobacter has been classified in species i.e., Azotobacter chroococcum, A beijerinckii and A. vinelandii. The first two species are deemed to be the most commonly occurring species in India. Azotobacter chroococcum appears in acid soils while A. beijerinckii in neutral and alkali soils. The beneficial effects of Azotobacter bio-fertilizer on cereals, millets, vegetables, cotton and Sugarcane, under both irrigated and rain-fed field conditions, have been susbtantiated and documented. Application of azotobacter has been found to increase the yields of wheat, rice, maize, pearl millet and sorghum by over 0.30% control. Apart from nitrogen, this organism is also capable of producing anti-bacterial and anti-fungal compounds, hormones and side rophores.

Azospirillum Inoculants

The ability to fix nitrogen by certain spirilla (Spirillum linoleum) was first recorded by Beijerinck in 1925 and later on confirmed by several workers using not only conventional micro-kjeldahal assay but also by the methods such as acetylene reduction technique and the isotopic enrichment method involving N. Tarrand and his associates (1978) renamed this organism as Azospirillum (N_2 fixing Spirillum). Certain micro-organism like bacteria and blue-green algae have the ability to use atmospheric nitrogen and transport this nutrient to the crop plants, while others colonize the root zones and fix nitrogen in loose association with plants. A very important bacterium of the latter category in Zaospirillum which was discovered by Brazilian scientist attracted head-lines in the mid 1970s. The crops which respond to Azospirillu inoculation are maize, barley, oats, sorghum, pearl millet and forage crops. Azospirillum applications increase grain productivity of cereals by 0.20% of millets by 30% and of fodder by over 50%.

Azolla-anabaena Symbiosis

Azolla is an aquatic fern commonly found floating in ponds, pools, tanks, shallow, ditches and channels. It is also fond in rice fields. Nitrogen fixing Blue-green algae (GBA) Anabaena azollae is found in the cavities on its dorsal leaves and fixes atmospheric nitrogen efficiently which is available to Azolla plants. The recognition of Azolla contains (0.2-0.3%)

nitrogen on fresh weight basis and 3-5% nitrogen on dry weight basis.

A small floating water fern, Azolla is a commonly seen in low land fields and in shallow fresh bodies. This fern harbours a blue-green algae, Anabaena azollae. The Azolla-Anabaena assocaition is a live floating nitrogen factory using energy from photosynthesis to fix atmospheric nitrogen currently used as a bio-fertilizer for rice in China, Vietnam, India, Indonesia, Thailand and other East and South Asian countries are available. An integrated system of rice Azolla fish has been developed in China.

Mycorrhizae: Special method for Obtainning Essential Elements

Mycorrhizae is the symbiotic association of fungi with roots of vascular plants. The main advantage of mycorrhizae to the host plants lies in the extension of the penetration zone of the root fungus system in the soil, facilitating an increased phosphorus uptake. The inter-connected net-work of external hyphen acts an additional catchments and absorbs surface in the soil beyond the dipletion zone that would otherwise be inaccessible to the plant roots. Endotrophic mycorrhizae has been shown to be present in wide range of horticultural species including apple, walnut, almonds, citrus, avocado, strawaberry and grape. In many cases the mycorrhizae have been shown to markedly improve the growth of the plants.

Phosphate Mobilizing Biofertilisers

The efficiency of phosphatic fertilizers is very low (15-20%) due to its fixation in soil. Besides, native soil phosphorus is mostly unavailable to crops because of its low solubility. The introduction of efficient P solubilizers in the rhizosphere has been found to increase the availability of phosphorus from both applied and native soil P the phosphorus which gets fixed in soil is made available to crops by the action of micro-organissms which solubilize insoluble/fixed forms of phosphates into forms which are readily taken up by the plants. Another factor which affects the effective ultilization of P is its low mobility in soil. Fortunately some filamentous organisms like vesicular carbuncular mycorrhizae (VAM) which form symbiotic association with plant roots have the ability to mobilize phosphorus from soil and thereby helping in absorption of P by plant roots. Commercial exploitation of phosphate bio-fertilizers can play an important role particularly in making the direct use of abundantly available low grade phosphate possible.

VERMI-COMPOST

It is a method of making compost with the use of earthworms which generally live in soil, eat bio-mass and excrete it in digested form. This compost is generally called vermin-compost or wormi-compost. It is estimated that 1800 worms which is an ideal population for one sq meteoric feed on 80 tonnes of humus per year. To prepare an ideal vermin-compost the following procedure is adopted:

Each shed measuring 20 ft 80 ft is to be constructed with the help of locally available material like bamboos, stems of trees etc. A hut type structure is built with the help of these articles. The roof is made from dried grass, typhus leaves, bamboo sticks, etc., in such a way that the hut may be protected from rain water and sun heat. Each hut may accommodate at least four vermin-beds of 3 ft. width.

These beds are prepared by putting 2 to 3 cm thick layer of farm manure as first layer followed by 10 15 cm of bio-mass with 200-250 worms per sq. ft. collected locally it may be added that the bed should be kept sufficiently moist. This layer should be followed by a layer of 10 15 cm of half digested cow-dung layer which should be covered by a layer of leaves, trash etc. and water is sprinkled on the entire bed. The bed may be covered with palm leaves or coconut leaves or with any indigenous material. The pits should be kept constantly moist but never flooded.

A month later, the covered leaves should be removed and layers of organic waste not exceeding 6-7 cm should be added every alternate day. Watering should continue with each filling. When the pit is nearly full to a height of one metre, the material should be turned to provide aeration. After a month, the heap will be ready for harvest with good quality vermin-compost. The dug out vermin-compost should be heaped in an open place. The worms will find way to the bottom of the heap. The vermin-compost from the top can be removed, dried and sieved for application in the field. The compost can also be enriched with micro-nutrients, bacteria etc. by adding it to them externally. About 16 tones of compost can be obtained from 4 beds in 30 days after four months period of gestation. (Courtesy Organic Farming - Arun K. Sharma)

TRAINING AND PRUNING

To avoid damage due to stem borer, a multi-stem training with 4-5 stems per hill was found beneficial. Removal of water shoots, cross branches, dead and diseased portions and suckers are necessary. Pruning of old spurs (short branches) to encourage growth of new ones is suggested. Some useful tips on pomegranate pruning are:

1) Fruitful and differentiated buds are located at the distal portion of the branches.
2) Pruning of terminal portion of branch would lower down the total flower production.
3) Pruning does not affect sex ratio and fruit quality in pomegranate,
4) Pruning significantly affects total fruits, marketable and unmarketable. Fruits and yield of higher grade fruits is more with intensity of pruning and
5) Pruning minimizes the bending and also the staking. Based on estimated money returns, 40 cm pruning of main stems was the best.

Nutrition

Nutritional studies (Chougule, 1976; Shende, 1997) indicated that following fertilizer schedule should be given:

Age (Year)	FYM (Kg)	Gram/tree/year		
		N	P	K
1	10	250	125	125
2	20	250	125	125
3	30	500	250	250
4 onwards	50	625	300	300

Full dose of FYM P and K and half dose of N should be given at bahar treatment prior to first irrigation. The remaining quantity of N can be given in single dose 40-45 days after the first dose. Both macro and micro-nutrients affect growth and development in pomegranate. The deficiency symptoms and leaf nutrient standards for macro-nutrients like N, P, K, Ca, Mg and S have been described by(Wavhal, 1986) in sand culture studies on folia applications of B, Mn Zn and Fe indicating beneficial results. Boron application at 0.2 per cent was found useful to reduce cracking of fruits and improvement of fruit colour. Lal (1975), Pundir and Pathak (1981) and Bankar et al., (1990) observed that soil application of NPK fertilizer was useful for increasing the production and fruit quality of pomegranate. Lal and Chauhan (1979) reported highest fruit yield with a medium leaf N content (2.35 per cent). Bhargava and Dander (1987) observed that most of the nutrients were stabilized in the 8th pair of leaf from growing tip in pomegranate and may be taken as standard from nutritional diagnosis. Mude et al., (1980) reported stable age of leaf for Fe, Mn and Zn as 4,5 and 11-12 months are respectively. (Courtesy Production Technology of Arid & Semi Arid Fruits - MPKV, Rahuri, 1996)

IRRIGATION

This crop can produce flowers and yields satisfactorily

even under conditions of low soil moisture and saline soils (Guseva,1960), but for optimum growth ample water is needed. Growth of plants was significantly reduced when soil moisture tension was greater than 2 atom. Observe that check basin requires 108 ha. cm of water, whereas drip with much required 40.31 ha. cm of water. Water saving was to the extent of 44 per cent in the drip system and 65 per cent when sugarcane trash was used in drip application of water equivalent to 20 per cent wetted area it is superior over surface method and average annual irrigation water requirement of pomegranate through drip method which is 20 cm. Saving in water to an extent of 43 per cent of increase yield (30-35 per cent as is reported by them.

It is observed that more number of flowers and fruits yield under 0.8 IW/CPE ratio, which was increased to 1.0 during flowering and fruit development and decreased to 0.6 during bahar treatment. It is reported that in drip trial water use efficiency to be maximum (3.38 q/ha cm) in 20 per cent wetted area daily treatment followed by 20 per cent wetted are every alternate day-93.32 q/ha cm.).

Effect of saline water

High density of soils and irrigation with saline water affects normal fruits production. It is reported that irrigation with 6.5 mm hos for 8.10 months did not show any adverse effect on survival of cultivar Khog. Jalore Seedless could tolerate up to 4.5 mm hos electrical conductivity.

Fruit growth and development

The maximum growth rate (2.09g/day) of the fruit is attained between 60-120 days stage in Ganesh and better fruit size can be attained by proper management during this period of active growth.

Water/moisture conserving measures

Black polythene followed by sawdust and banana trash proved to be the better material (Chattapadhyay et al., 1992)

10 per cent kolin, 1.5 percent power oil and one per cent paraffin were found beneficial to increase productivity (Avon, 1995).

PHYSIOLOGICAL DISORDERS

Internal breakdown of fruits

Unfortunately the production of good quality fruits of many varieties is be set by development of serious malady known as internal breakdown or internal blackening, wherein, externally the fruit appears quite normal but its arils become soft having light creamy brown to dark, blackish brown colour. The incidence of internal breakdown after 150 days of an thesis and its intensity increases if the fruits are left on the tree up to 165 days. The incidence of browning increases with increase in weight of fruit. It is reported that TSS, acidity, ascorbic acid, total sugars, reducing sugars, calcium, phosphorus and the enzyme catalase were low whereas no-reducing sugars, starch, tannins, nitrogen, potassium, magnesium, boron, polyphone oxidize and peroxides enzymes were high in affected arils than the healthy ones.

Cracking of fruits

Cracking of fruits are believed to be due to a wide variation in soil moisture content. Accordingly high air temperature rise was found to be the cause of cracking. It amounted to 63 per cent in spring crop Jan-June, 34 per cent in winter crop (Oct-March) and only 9.5 per cent in rainy season crop (July-Dec) in variety Jodhapuri. It is observed that cracking of fruits as a varietals character and none of the cultivars was absolutely free of this trouble. Rind thickness and texture in various cultivars seem to be related to cracking. Cracked fruit shows reduction in weight of fruit and grains and volume of juice per fruit.

PESTICIDE CONTROL SYSTEM

Most countries have brought out legislation and

maximum residue limits have been specified for various horticultural produce including pomegranate. For ensuring pesticide control, it is essential to have an effective control system and assess the effectiveness of the same. The following measures are essential for having an effective pesticide residue control system:

1. A pesticide usage programme of the producer shall be submitted prior to production for each season.
2. A field pesticide application record shall be maintained by the producer.
3. Pesticide residue limits as specified by the importing country shall be maintained. Pomegranates for export to a certain country should be tested for limits specified by that country. All consignments of pomegranates for exports to the European Union shall be compulsorily tested for pesticide residues. These tests shall be conducted for those chemicals which are either banned or where there is a possilility of having higher residue levels that those permitted a list of chemicals requires compulsory testing for export to the Eurpopean Union.
4. Routine random pesticide residue analysis shall be carried out as per the guidelines given in Guide in Codex Recommendations Concerning Pesticide Residues; Part 5 Recommended Method of Sampling for the Determination of Pesticide Residues.
5. In case of high residue levels, a plan needs to be drawn up to investigate into the causes and take corrective action to prevent future occurrences. This should include:

a. An investigation into the possible causes of high residue level.

b. Re-sampling or repeat of the residue analysis with, if possible, an additional independent analysis.

c. Withdrawal from the sale or distribution of product from which the sample was taken and its isolation from any other product until the product has been declared safe.

d. Informing the appropriate authorities and customers of the high residue levels as soon as the same is detected if the product is in the marketing chain.

6. The international, European and Indian limits for pesticide residues are given in the following publications:

International

- Codex Alimentarios Volume 2 and 2B (1996) Pesticide Residues in Food (Second Edition), 2000
- Codex Alimentarius Volume 14 Acceptances

European

- Council Directive 76/642/EEC of 23rd November., 1976 realing to the fixing of maximum levels for pesticide residues in and on fruit and vegetables.
- Coucil Directive 90/642/EEC of 27th November, 1990 on the maximum levels for the pesticides residues in and on certain products of plant origins, including fruit and vegetables.
- Council Directive 93/68/EEC of 29th June, 1993 amending Annex, 11 to Directive 76/895/EEC relating to the fixing of maximum levels for pesticide residues in and on fruit and vegetables and the Annex to Directive 90/642/EEC relating to the fixing of maximum levels for pesticide residues in an on certain products of plant origin, including fruit and vegetables, and providing for the establishment of a first list of maximum levels.
- Councial Directive 94/30/EC of 23rd June, 1994 amending Annex, 11 and Directive 90/646/EEC relating to the fixing of maximum levels for pesticide residues in and on certain products of plant origin, including fruit and vegetables and providing for the establishment of a list of maximum levels.

- Council Directive 96/32/EEC of 21st May, 1996 amending Annex 11 to Directive 76/895/EEC relating to fixing of maximum levels of pesticide resides in and on fruit and vegetables and Annex 11 to Directive 90/642/EEC relating to the fixing of maximum levels of pesticide residues in an on certain products of plant origin, including fruit and vegetables and providing for the establishment of list of maximum levels.
- Council Directive 97/41/EC of 25th June, 1997 amending Directives 76/895/EEX, 86/362/EEC, 86/363/EEC and 90/642/EEC relating to the fixing of maximum levels for pesticide residues in and on, cereals, foodstuffs of animal origin and certain products of plant origin, including fruits and vegetables.
- Council Directive 2000/24/EC of 28th April, 2000 amending the annexes to Coucil Directives 76/895/EEC, 86/363/EEC and 90/642/EEC on fixing the maximum levels of pesticide residues in and on cereals, food stuffs of animal origin and certain products and plant origin including fruits and vegetables.

Indian

- Fruits and Vegetable Grading and Marking Rules, 2002.
- Prevention of Food Adulteration Act, 1954 and Rules Revised 1998 framed thereunder.

CHEMICALS REQUIRING COMPULSORY TESTING FOR EXPORT TO EUROPEAN UNION

A. MERCURY COMPOUNDS

1. Mercuric oxide
2. Merucurous chloride (calomel)
3. Other inorganic mecury compounds
4. Alkyl mercury compounds
5. Aloxylalkyl+aryl mercury compounds

B PERSISTENT ORGANO-CHLORINE COMPOUNDS

1. Aldrin
2. Chlordane
3. Dieldrin
4. DDT
5. Endrin
6. HCH containing less than 99% of the gamma isomer
7. Heptachlor
8. Hexachlorobenzene
9. Comphechlor

C. OTHER COMPOUNDS

1. Banned substances
   a. Ethylene oxide
   b. Nitrogen
   c. 1,2-dibromoethane
   d. 1,2-dichloroethane
   e. dinoseb (acetates and salts
   f. Binapacryl
   g. Captafol
2. The substances below are not banned if they meet purity requirements stated against each:
   a. Dicofol; Products containing more than 78% of p.p. dicofol and less than 1g/kg DDT related compounds.
   b. Maleci hydrazide; Choline, potassium and sodium salts containing less than 1 mg/kg free hydrazine (acid equivalent)

c. Quintozene Products containing less than 1g/kg HCB and less than 10 g/ka petachlorobenzene.

MAXIMUM PERMITTED PESTICIDE RESIDUE LEVELS FOR POMEGRANATE FRUITS-EC, CODEX AND INDIAN

Pesticides/Chemicals	Limits (mg/kg)		
	EC	CODEX	Indian
Acephate	0.02		
Aldrin and Dieldrin		0.10	
Aminotriazole	0.05		
Aramite	0.01		
Atrazine	0.10		
Azenphos methyl		1.00	
Barbon	0.05		
Benefurocarb	0.05		
Binapacxyl	0.05		
Benomyl			5.00
Binapacryl	0.05		
Bromide ion		20.00	30.00
Bromophos ethyl	0.05		
Camphechlor	0.10		
Captafol	0.02		
Captan			15.00
Carbendazim			5.00
Carbofurna	0.10		0.10
Carbosulfan	0.05		
Chlorbenside	0.01		

Chlorbufam	0.05		
Chlordane		0.02*	0.10
Chlorobenzilate	0.02		0.10
Chlorpyrifos			0.50
Chloropyrifos methyl	0.05		
Chlorothaloni	0.01		
Chloroxuron	0.05		
Copper oxychloride			20.00
Cyhalothrin-lambada	0.02		
cyfluthrin	0.02		
Cypermenthrin	0.05		
D.T.T.	0.05		3.50
Daminozide	0.02		
Diallate	0.05		
Diazinon	0.02		
Dichlorprop	0.05		
Dichlorvos			0.10
Dicofol			5.00
Dimethoate			2.00
Dinoseb	0.05		
Dioxathion	0.05		
Disulfoton	0.02		
Dithiocarbamates			3.00
Endosulfan		2.00	2.00

Endrin	0.01		
Ethion			2.00
Ethylene dibromide	0.01		
Fenarimol	0.02		
Fenchlorphos	0.01		
Fenitrothion			0.50
Fentin	0.05		
Fenvalerate	0.05		
Formathion			1.00
Furathiocarb	0.05		
Heptachlor	0.01		
Hexachlorocyclo hexane(all isomers)			1.00
Fentin	0.05		
Fenvalerate	0.05		
Formathion			1.00
Furathiocarb	0.05		
Heptachlor	0.01		
Hexachlorocyclo hexane (all isomers)			1.00
Lindaane			1.00
Malathion			4.00
Maleic hydrazide	1.00		
Maneb or Mancozeb	0.05		
Mehamidophos	0.01		
Mehoxychlor	0.01		
Methylbromide	0.05		
Monocrotophos			1.00
Paraquat	0.05		0.05
Parathion			0.50
Parathion methyl			0.20

Phorate	0.05		
Phosalone			5.00
Phosphamidon			0.20
Propyzamide	0.02		
Pyrethrins			1.00
TEPP	0.01		
Thiometon			0.50
Trichlorofen			0.10
Triforine	0.05		
Vinchlozolin	0.05		
2,4-D			2.00
2,4-T	0.05		

PACKING AND LABELLING

Packing

Fresh pomegranates shall be packed in such a way that the produce is suitably protected during transportation and handling. The containers shall meet the quality, hygiene, ventilation and resistance characteristics to ensure suitable handling, shipping and preserving of the pomegranates. Packges or lot must be free of all foreign material and smell.

The container shall comprise of cardboard cartons of the inter-locking type preferable having a water-proof to prevent damage due to high humidity in cold stores.

The mechanical strength of the package should be sufficient to survive pre-cooling/cold storage procedures for the ability to withstand stacking without getting distorted or collapsing during the normal storage life under the storage conditions specified.

The cartons shall be well ventilated and provided with

round holes for adequate ventilation.

Specifications details for corrugated fiber board boxes for packaging pomegranate for export as suggested. Boxes conforming to these requirements should be used and which have been treated with wax. Fruits should be packed in cardboard boxes which should be slotted.

The material used for the inside should be new, clean and of a quality such as to avoid causing any external or internal damage to the produce. The use of materials and particularly of papers or bearing trade specifications is allowed provided that printing or labelling has been done with a non-toxic ink or glue.

While packing fresh pomegranates for exports, the packing requirement of the importing countries should also be taken into account.

Palletisation: Pallets of dimensions 1000 mm x 1200 mm and 800 mm x 1200 mm (only for exports to Europe) should be used. Pallets should be strong enough to ensure transit till the final destination. These should be secured by plastic or fiber corner boards. Strapping and clips used should preferably be of plastic material.

Presentation: In addition to the requirements given above, the criteria for presentation as given below shall also be followed in case of exports to the European Union.

a. The contents of each package must be uniform, each package must contain pomegranates of the same origin, variety, class and degree of ripeness.

b. The visible part of the contents of each package must be representative of entire contents.

Labelling

Each package shall bear the following particulars in letters grouped on the same side, legibly and indelibly marked and visible from the outside.

a. **Identification**

Packer; Name and address or official issued or

And/or; accepted code mark

Dispatcher

b. **Nature of produce**

Pomegranates if the contents are not visible from the outside.

Name of the variety

c. **Origin of produce**

Country of origin and optionally district where grown, ornational, regional or local place name.

d. **Commerical specification**

Class

e. **Official control mark (optional)**

I. Net weight

II. Data code

All information shall be on the same side of the package. The marking may be done by using an ink stamp, by printing on to a package by means of a label firmly fixed to the package or by a combination of these methods.

It is recommended that the information is marked on both ends of the package it is rectangular or on the two opposite sides, if it is square.

The recommended storage temperature may also preferably be printed on the package.

In addition to the requirements given above, the labelling requirements of the importer shall also be taken into account.

SPECIFICATIONS FOR CORRUGATED FIBREBOARD BOXES FOR PACKING POMEGRANATES FOR EXPORTS

Usually for packing pomegranates for export purposes a cardboard corrugated fiberboard box of 4.0 or 5.0 kg capacity is used. The dimensions of such boxes depending upon the capacity are:

i. 375 x 275 x 100 mm

ii. 480 x 300 x 100 mm

The detailed specifications are given below:

Sl No.	Specifications	Ring and Flap Tuch in type	RSC Regular Slootted Container	Slide Type
1	Material for construction	5 ply	5 ply	5 ply
		CFB	CFB	CFB
2	Gram mage Gm/m^2 outer to inner	* 230 x 140 x 120 x 140	*230x 40 x140x 140	*230 x 140 x 140 x 140
3	Bursting strength Kg/cm^2	Min 10.00	Min 10.00	Min 10.00
4	Puncture resistance Inches/teat inch	Min 250	Min 250	Min 250
5	Compression strength kgf"	Min 350	Min 350	Min 350
6	Cobb value (30 ming/m^2)	Max 30	Max 130	Max130

* Outer ply of white duplex board

MAINTENANCE OF RECORDS

Record shall be clearly written, dated and signed by the person entering the data. In case of any changes in the record, these shall be signed by the authorized person. All records shall be filed carefully and be retrievable and available for inspection at any time by any person authorized to have access to the records.

Time period for storage of various records should be clearly laid down. The length of storage period depends on importance of the record relative to the product. No record should be destroyed within the period of a product being marketed.

LIST OF ESSENTIAL RECORDS

Sl No.	Record	Time period
1	Management review records	2 years
2	Customer order records	1 years
3	Records of contract review	1 year
4	Records of approved producers and other suppliers	1 year
5	Records of raw materials-conditions supply agreement	3 years
6	Final field survey record	6 months
7	Raw material inspection records	6 months
8	Raw material quality control report	6 months
9	Equipment inspection records	6 months
10	Production records	6 months
11	Final product in inspection records	6 months
12	Calibration records	3 years
13	Pesticide application data	1 year
14	Pesticide residue records	3 year
15	Training records	As per policy
16	Internal quality audit records	3 years

SECTION AND TRAINING

A systematic and well structured procedure should be developed for identifying training needs and providing for training of farmers, exporters and pack-house personnel.

Training in the following areas is recommended:

FOR FARMERS AND FIELD STAFF OF EXPORTERS OF POMEGRANATES

a. For producing export worthy pomegranate of good edible quality, free of diseases, insect injury, free blemishes and also for proper maintenance of an orchard from hygienic point of view.

b. Use, storage and application of pesticides.

c. Techniques of pesticide residue analysis.

FOR FIELD STAFF AND PACK-HOUSE PERSONNEL

a. Raw material assessment

b. Final and In-process assessment

c. Processing operation including washing, waxing, fungicide application, drying, preparation of disinfectant, fungicide and wax solution, etc., safe handling techniques of chemicals and probability of contamination and microbial growth.

d. Sizing and grading

e. Packing methodology

f. Storage conditions of fresh pomegranates

g. Hygienic requirements-techniques of hygienic handling and storage of fresh pomegranates.

h. Awareness and responsibilities—personnel associated with packing should be aware of good management practices, good hygienic practices and their role and responsiblity in protecting fresh pomegranates from contamination and deterioration. Packers should have necessary knowledge and skills that minimizes the potential for microbial, chemical and physical contamination.

i. Use of equipment and methodology of quality analysis.

j. Equipment calibration

Training records shall be maintained at both individual and organizational level.

DOCUMENT AND DATA CONTROL

Documents and data including standards, procedures, work instructions, proforma, etc. shall be under control. The control should include the following:

a. All documents requiring control shall be identified and a master list of these shall be prepared and be readily available.

b. The location for each documents shall be identified and current relevant issues of these should be available at identified points.

c. Obsolete documents should be removed from all points of use.

d. Changes made in documents should be approved by the nominated authorized person.

Confidentiality—Many files such as those containing customers specifications are of confidential nature. The confidentiality of these shall be maintained. The supervisors or workers may be issued an outline specification and the original kept securely.

A list of documents which need to be under control is given below:

- Company manual and procedures
- Customers specifications
- Legislation of importing country
    - Pesticide residues, heavy and metal limits toxic residues, fertilizers application, etc.

- labelling requirements
- Weight/volume requirements

- Cleaning schedule for pack-house
- In-process quality control worksheet
- Calibration methods for various equipment

SECTION AND CODE OF HYGIENIC PRACTICES FOR POST-HARVEST HANDLING OF FRESH POMEGRANATES FOR EXPORT

Due to year-round availability of fresh fruits from a global market, there is a substantial increase in consumption of fresh fruits and vegetables in the past two decades. Recent increase in report of food-borne illness associated with fresh fruits has raised concerns from public health agencies and consumers about the safety of such farm products.

It is with this background, we are dealing with Code of Hygienic Practices both at pre-harvest i.e. primary production and at post-harvest handling stage. The Code of Hygienic Practices concerning primary production of fresh pomegranates has been covered in the pre-harvest manual for production of fresh pomegranates for export.

Code of Hygienic Practices at post-harvest processing, handling and packing stage are discussed in this chapter.

Packing establishment; DESIGN AND FACILITIES

Hygienic aspects in pack house design etc. have been covered in Section 2 of this Manual i.e., Process control and management of quality during production: All aspects concerning hygiene have been discussed under sub-section 2.2 pack house requirements.

Control of operation

To avoid contamination from microbial and chemical hazards, entire operations i.e., both at pre-harvest as well as at post-harvest level have to be monitored. Every possible food hazard which can cause contamination of fresh pomegranates has been considered at pre-harvest stage and possible preventive and control systems have been described in chapter 2 of manual for production of fresh pomegranates for export. At post-harvest stage possible hazards which can cause contamination are described in the next paragraphs.

Key aspects of hygiene control systems

- The details of temperature control system for maintaining hygiene and producing wholesome fresh pomegranates have been described in section 2 process control and management of quality during production.

Post-harvest water use

- Packers should follow good management practices to prevent or minimize the potential for introduction or spread of pathogens in the prcessing water. As far as possible only potable quality of water should be used. Guidelines for hygienic re-use of processing water are as follow:
- Post-harvest systems that use water should be designed in a way that product does not lodge and dirt should not be allowed to be built up.
- Disinfectants should only be used where absolutely necessary. The disinfectant levels should be monitored and controlled to be maintained. After disinfection, washing with clean water is absolutely essential to ensure that chemical residues don't exceed levels as recommended by Codex Alimentations Commission.
- Where necessary, the temperature of post-harvest water should be controlled and monitored.
- Recycled water should be treated and maintained in a way that it does not constitute a risk to the safety of fresh pomegranates. The treatment process should be

effectively monitored and controlled.

Chemicals treatments

- Packers should only use chemicals for post-harvest treatments (waxes and fungicides) in accordance with the general standards on Food Additives or with the Codex Pesticide Guidelines. These treatments should be carried out in accordance with the recommendations for the purpose.
- Sprayers or traversing automatic nozzles for post-harvest treatments should be calibrated properly to control the accuracy of rate of application. These should be throughly washed when used with different chemicals and different fruits to avoid contamination.

Cooling of fresh pomegranates

- Forced air cooling in case of fresh pomegranates is the use of rapid movement of refrigerated air over fresh pomegranates in cold rooms. The air cooling system should be appropriately designed and maintained to avoid contamination of fresh produce.

Cold storage

- When appropriate, fresh pomegranates should be maintained at low temperatures at recommended levels after pre-cooling to minimize microbial growth, the temperature of the cold storage should be controlled and monitored.
- Condensate and defrost water from the cooling system in cold storage area should not drip on to fresh pomegranates.
- Inside of the cooling system should be maintained in a clean and sanitary condition.

Packing

- Hygienic aspects during packing of fresh pomegranates

have been fully covered under section 5 of this manual.

Packing establishment-maintenance and sanitation

Maintenance and sanitation of equipment and pack-house hygiene has been fully covered in sub-section 2.3.1. Equipment and pack-house hygiene under section 2 process control and management of quality during production of this manual.

Cleaning schedule of pack-house for pomegranate has also been provided.

Personnel hygiene

All aspects of personnel hygiene in the pack-house have been covered in section 2 as mentioned above, in sub-section 2.32 and 2.4 of this manual.

Transportation

All hygienic aspects during transportation of fresh pomegranates in conveyances or containers have been covered in sub-section 2.5 transportation requirements under section 2 of this manual.

Training

All aspect of training covering awareness and responsibilities, hygienic requirements and training needs of pack-house personnel have been covered in section 6 of this manual.

Documentation and records

Maintenance of Records, Document and Data control have been adequately covered in section 56 and 7 respectively. However, supplementary information is enumerated below:

- Where appropriate records of processing, production etc., should be kept long enough to facilitate trace back food borne illness if investigation is required. The period of holding back records could be longer that shelf life of fresh pomegranates. Documentation can enhance the credibility and effectiveness of food safety control system.

- Growers should keep all relevant information on agricultural activities like site of production, information on agricultural inputs given by suppliers, lot numbers of agricultural inputs, use of pesticides, fungicides, water quality data, cleaning schedules of equipment, containers etc.
- Packers should keep all information concerning each lot such as inforamtion on incoming material (information from growers,), data on quality of processing water, fungicide and application programmes, cooling and storage temperatures and cleaning schedules for pack-house premises, packing line, equipment, containers etc.

Re-call procedures and trace back

when appropriate growers and packers should have trace back programmes to ensure effective lot identification, the system should be able to trace the sites and agricultural inputs involved in primany production and the origin of incoming material at the pack house in case of suspected contamination.

Grower's information should be linked with packer's information so that the system can trace products from distrubutor to the filed. Information that should be included are date of harvest, farm identification and where possible, the persons who handled the fresh fruits (pomegranates) from the primany production site to the packing establishment.

SECTION-9; INTERNAL QUALITY AUDITS

An internal audit programme shall be established by the pack-house to cover the entire system, facilities and operations. The nature of the programme will depend on the size of operations.

A programme of time-table of audits covering all systems be drawn up. Audits shall be carried out by the management staff not directly involved with that activity.

A suggested programme of internal audits is as give below:

Sl. No.	**Area of Audit**	**Frequency**
1	Pack-house audit	Once every 3 months (a sample audit sheet)
2	Quality Assurance Audit	Once in 6 months (minimum)
3	Audit of entire operation the different areas of operation production till the final package in dispatch of the product.	Once a year (minimum)
4	Maintenance of files and records (to cover maintenance of up of suppliers files, specification, rejection, visit record, etc.)	Once a year (minimum)

Records of the audit shall be maintained for a minimum period of 3 years.

SAMPLE AUDIT SHEET FOR PACK-HOUSE AUDIT

Location ______________ Audit team 1) ______________

2) ______________

Date of previous audit __________ 3) ______________

Sl No.	Particulars	Action
		By whom, By when
1	**Fabric of building** Repairs required (if any) Improvements	

Sl No	Particulars	Action
		By whom, By when
2	**Vermin and bird control** Entry points—doors, windows, etc. routine controls - Baits in place and numbered - Visit report in file - Baiting plan in file	
3	Contamination/hygiene risks Toilets - Cleaning schedule in force - Hand-wash, soap and hand—dry facilities in force Pack-house - Cleaning schedule in force - Clearance of rubbish - Storage of packaging and chemicals - Water quality - Personal hygiene in force - Work instructions in place - Pack-house disciples in force Glass - Lights protected, etc.	
4	**Equipment** - Maintenance and condition - Repairs, improvements - Weight checks, records - Calibration records	

Report agreed and signed by: 1)______________

2) ______________

3) ______________

Date ______________

Copies to relevant managers for action ________

SUMMARY OF RECOMMENDATIONS AND GUIDELINES TO BE FOLLOWED FOR POST-HARVEST HANDLING OF POMEGRANATES TO ENSURE QUALITY

Usually, in India post-harvest handling not been taken very seriously. This aspect is very technical and, therefore following methodology needs to be adopted in post-harvest handling of pomegranates:

Selection of pomegranate orchard

It is always best to select an orchard where recommended fungicide, pesticide and balanced nutrients have been applied at the required and appropriate time. Following guidelines should be considered while selecting a pomegranate orchard for expert purposes:

- An orchard where last pesticide application has been made at least one month before harvest.
- The orchard should be apparently healthy, free of nutrient deficiencies, free of sunscald, orchard clean and not strewn with fungal or bacterial infected leaves, fruit borer infested fruits and also free of rusting (mite damage) and rough surface on fruits (thrips damage).
- Fruits should be of adequate size and diameter characteristic of the variety.
- Ensure that the fruits have reached the required maturity and colour. The brilliant fruit shine and appearance is most important in terms of value realization foreign countries. This must be looked into while selecting the orchard.

- Some tests are necessary at the field level to find out quality parameters of fruits while selecting the pomegranate orchards. The tests to be done and the requirements are given below:

i) Weight	-	Weighing scales
ii) Size of fruits	-	Plastic rings of different sizes
iii) T.S.S.	-	Refract meter
iv) Temperature of fruit	-	Temperature probe

Maturity and maturity standards

It is generally considered that the fruit is ready for harvest when fruits have attained the distinctive fruit colour characteristic of the variety and make a metallic sound on tapping.

Maturity standards vary with the cultivar. For example, Ganesh cultivar is considered mature when T.S.S. reaches 14.50 B or so and acididty % comes to a level of 0.3% or 0.35%. The maturity indices vary slightly for Mridual cultivar as in this if T.S.S. has reached 16° B or so and acidity around 0.45% it is considered fir for harvest.

Quality requirements of importing countries

In foreign countries especially in European countries, where it's demand is maximum like brilliant shiny crimson or red colour of fruits and sparkling red colour of arils (kernels). This is because whole fruits are widely used for decoration and kernels are used for garnishing desserts and salads. Also juice from dark-red kernels is used for syrup making which is further used for flavouring mixed drinks, topping of ice cream and desserts. Therefore, desirable fruit characters of importing countries are enumerated below:

i) Smooth cutting at the stem end

ii) Shining dark-rose pink colour of the fruit

iii) Fruit weight around 200-400 gms,

iv) Round shape of fruit,

v) Uniform size and shape of fruit in pack,

vi) Dark rose pink arils (kernels),

vii) Softness of seeds,

viii) High total soluble solids near about 16° b to 17°B,

xi) Freedom from scars, belmish,es scratches, rusting, rough surface, disease spots and fruit borer,

x) Bracts of calyx having freshness and without any damage,

xi) Pleasant flavour and aroma and

xii) Without blackening of arils.

Method and time of harvesting

- Harvest pomegranates either early in the morning or in the evening hours,
- Use clippers and bags for harvesting,
- Do not pull the fruits,
- Cut the stem and close to the base,
- Do not bruise the base while cutting the fruits. Handle carefully at harvest,
- Do not keep longer stem end, as it can cause damage to other fruits,
- Keep the fruits in plastic crates for emptying the bags,
- Do not keep fruits on the ground,
- Do not keep empty creates in sun but in shade,
- Do not overfill the crates with fruits and keep them in shade and not in open sun,
- Do not damage calyx while harvesting or while keeping in crates or during field storage,

- Crates should be washed with water and detergent and Clippers should also be cleaned frequently with wet cloth and dipped in 1% sodium hypochlorite sol.

Transporting fruit from field to pack-house

- Transport the fruits immediately to pack-house without any loss of time,
- Transport should be available on demand to avoid any time lag between harvest and transport to pack-house
- Shiney appearance and bright color of rind are prime factors for higher value realization and therefore shine and appearance of fruits should not be spoiled during loading unloading and during transport,
- There should be no jerks or bumps in the vehicle to cause rubbing among the fruits,
- The vehicle should be clean to avoid infection to the fruits and
- Preferably no other product should be tranported along with pomegranates.

Receipt of pomegranates at pack-house

- On arrival, the unloading should be done gently to avoid bruising and preferably in the sorting bay having cool temperature to avoid multiple handling of the fruits.
- Do not stack too many crates over each other to avoid tumbling and bruising of the fruits
- Note down both fare and net weights of crates.
- Care needs to be taken that grower's code is maintained and produce from different growers does not mix.

Sorting, washing and treatment with fungicide and wax on a packing line.

i) Sorting

Pomegranate fruits in plastic crates are directly put under

the inspection roller-cum-conveyer. At this point, all fruits showing greenish colour, fruits, with sunspots, affected by fruit borer or having disease spots, rusting and rough surface which cannot be accepted in the international market must be eliminated. Also fruits having cuts, blemishes, damages, etc. need to be sorted out manually and rejected.

ii) Washing

After rejecting all unwanted fruits, these must be washed. It is preferable to add a disinfectant like sodium hypochlorite @ 100-200 ppm. After chlorine disinfection, rinsing with clean water is a must. Only potable water should be used for all post-harvest processing purpose.

ii) Elimination of excess water on the fruit surface

After washing, excess water on the fruit surface must be removed. Efficiency of wax and fungicide application depends upon how effectively excess water adhering to fruits skin has been eliminated. This is usually done by making use of absorbent rollers.

iv) Waxing and fungicide application

Most common and effective fungicides are Thiabendazole or Benomyl at 1000 ppm concentration. The concentration of the fungicide will vary with the speed with which fruit passes on the conveyer and the quantity of fungicide deposited on the fruit surface by traversing nozzle. Each packing line has its own speed depending upon design, capacity and manufacturer and thus concentration of fungicide must be calibrated/worked out so that international residue limits are not crossed and there is sufficient deposit to protect from post-harvest losses.

Correct strength of the fungicide is absolutely important to avoid post-harvest losses. It is beneficial to prepare fresh fungicide solutions everyday just before application for maximum effectiveness.

a) Poor quality wax should not be used. Only good quality

edible wax needs to be used.

b) Monitor edidle wax quantity required on packing line for very good shine on fruit surface.

c) Monitor concentration of the fungicide which will have minimum residue on the frui and minimize post-harvest losses.

d) Effective wax coating is very important especially in pomegranates in shrivelling of the fruit. In such a situation, realization value or market acceptability is reduced.

e) While using readymade wax and fungicide solutions, their manufacturing date must be considered.

Drying the wax coating

The wax coating on the fruits is usually dried in suitable drying tunnel of the packing line at a temperature of about 30-35°C. However, it must be ensured that:

- Bulbs, heaters, hot air circulation system along with fans, air knives, etc. must work properly to attain proper drying.
- The temperature inside the tunnel must not go beyond 30-35°C

Sizing the fruits

After drying, the fruits are fed into feeder channels of a grader or size which works on the principle of roller sizing which is quite common in grading of fruits and gives good results. In this method, the distance between the rollers can be adjusted for grading different-sized fruits.

For pomegranate following grades are adopted:

For export purposes mostly 200 to 400 gm fruit weight are used. Weight of 400 to 450 gm and 450 to 500 gm are called 'A' grade fruits which are marketed within India in domestic market. Following steps must be taken in sizing of fruits:

Sl No.	Different size ranges	No. of fruits in	
		4.0 kg box	5.0 kg box
i.	45 to 500	Not exported	Not exported
ii	400 to 500	-do-	Do-
iii	350 to 400	10-12	12-15
iv	300 to 350	12-15	15-18
v	250 to 300	12-15	15-18
v	200 to 250	15-18	18-20

i) Variation in size grade should be minimum as far as possible,

ii) Quality supervisors must ensure that bad quality fruits must not go into the fruits which are to be packed in cardboard boxes.

Nose cleaning

Before packing a box, fruits at calyx and or calyx side are wiped with a cloth soaked in fungicide solution (Benomyl or Thiabendzole 1000 ppm) to avoid cross contamination in packed box.

Packing and coding

a. Packing methodology

For packing mostly 4.0 to 5.0 kg box is used and fruits 10 to 20 in number are packed in a box depending upon the fruit size. Usually a 5.0 kg box contains 10-15 fruits of Ganesh cultivar, whereas fruits of Mridula or Arakta 19-20 in number are accommodated. The diamensions of 4.0 kg box are 375 x 275 x 100 mm and that of 5.0 kg box are 480-300 x 100 mm. For Gulf countries paper shreds below the fruits in between and on the top of the fruits are used in packing to counteract any bruising during transit. However, for European market, cardboard partitions are used in packing to avoid movement of fruits within the pack and thereby bruising. To further safeguard the fruits from injury, bubble plastic is used and placed at the bottom, sides and at the top of the fruits.

The size of the box can be designed according to the requirements and accommodating a particular fruit-size which is liked in the importing country. However following suggestions must be taken into consideration:

(i) Cardboard boxes must have ventilation holes, hand holes, etc.

(ii) The packing should be such that there should not be pressing of any fruit so as to cause burns on the skin.

(iii) Placement of fruits by putting accurate-sized fruits should be such that there is no movement of fruits inside the pack.

(iv) The pallets should be dimensions namely 1000 mm x 12000 mm. For exports to Europe, pallets of dimension 800 mm x 1000 mm should be used.

(v) Only plastic material should be used for strapping and clipping for securing the cartons tc the pallets.

Boxes should be coded for date of packing, product's name, grower's code for tracing its origin. The box must be labelled with a very attractive label and should contain following information:

- Name of the produce
- Variety
- Origin of produce
- Commercial specification of class
- Date of packing
- Name of packing organization
- Gross weight/net weight

- Number of fruits/fruit size
- Recommended storage temperature
- Additional labelling requirement of importer should also be included.

Pomegranates should be chilled to 5-7°C temperature and at least within 10 hours after harvesting.

Before storing at above temperatures, fruits needs to be pre-cooled by forced air. Pre-cooling is very important for successful sea shipping. The ideal pre-cooling temperatures are the same, i.e., 5-7°C. There can be chilling damage if fruit is placed below 5°C for long time. The humidity should be around 90-95% during pre-cooling.

Following precautions may be taken during pre-cooling:

- Prompt pre-cooling is very important for avoiding post-harvest losses and distant transportation.
- The filled cartons should be shifted to pre-cooling chamber without any delay.
- On transfer of the product, the outflow temperature and humidity should be properly recorded.
- The products should remain in the pre-cooling chamber till the temperature of the fruit reaches to the required stage. Do not allow fall in temperature otherwise chilling injury will make the fruits unfit for consumption.
- The temperature, humidity and air circulation of the room shall be properly maintained and checked at pre-determined intervals.
- After pre-cooling and Palletisation, strapping should be done. This needs to be done only in coid room to maintain the cool climate.

Cold storage

The ideal storage temperatures are 5-7°C. The above

temperatures need to be kept throughout the transport (for destination to importing country) and further stored there. The temperature should never go below the figures stated, otherwise it is likely to result in chilling injury.

The relative humidity should be kept at 90-95% throughout transport and storage.

Susceptibility to chilling also varies with maturity and variety. Pomegranates subjected to too low temperature, develop brown discolouration of skin, get pale in colour of arils and brown discoloration of white segments separate the arils.

It is critical for pomegranate storage that cold storage should be uninterrupted. If there is interruption, it negates all prior careful handling of fresh product.

Quality appraisal

Quality checking should be done with regard to following parameters:

- Cutting stem end to appropriate size,
- Fruit size/uniformity,
- Colour of the fruit/uniformity,
- Condition of calyx,
- Freedom from sunscald, disease spots, fruit borer, rusting and rough surface,
- Fungicide residue in no case should exceeed prescribed limits,
- T.S.S. (°Brix) and
- Spoilage at the port due to diseases before dispatch to importing countries.

For ensuring export quality fruits, quality checking is suggested at following levels:

- Pack-house before dispatch to the port

- At the port before dispatch to the importing country
- This feedback is very important for further improvement. Acceptability and grading given in the importing country is very important for managing future exports.

Equipment required for quality control

- Refractometer - T.S.S.
- Burette, pipette, volumetric flask, glass rod, NaOH, Phenolphthalein for acidity analysis
- Scales for recording weight
- Pressure taster to test the firmness of the fruit
- Temperature probe to record temperature inside the fruit
- Spectrophotometer/HPLC to determine residue levels* of fungicides in the fruit.

Container loading

- In order to maintain the cool chain the temperature of the loading bay should be controlled preferably to 20-25°C
- The transfer of pomegranate from the cold store to the loading bay and to the container should be smooth and quick so that it does not result in temperature increase of pomegranates.
- Before shifting of pomegranates of containers of Reefer vans, ensure the appropriate recommended temperature and humidity inside the container.
- The doors of the containers must be closoed before and after loading pomegranates.
- Ensure that temperature and humidity recorders are placed inside the container.
- The cartons/pallets in the container are to be arranged in such a manner that, there is no movement within the

container. This can be done by placing the pallets in an interlocking manner or by us of iron roads/pegs for plugging the empty space.

- It is preferable or rather should be ensured that only a single type of product is loaded to one container.

Shelf life experiments

- A few cartons should be kept aside to check the shelf life of the product and should be tested periodically to ensure that the quality is maintained till it reaches the customer/ consumers. Separate areas in pack-house/cold store shall be set aside for the shelf life experiment boxes.
- Implementation of code of hygienic practice of post-harvest handling stage of fresh pomegranates.
- Design and facilities of pack-house should be such that there is no ingress of vermin from birds, etc. It should be protected by fence and flooring etc., to enable proper cleaning.
- Water for washing especially rinsing, should be of potable quality to minimize potential for introduction of pathogens, etc.

  Disinfectants wherever necessary should be monitored and residues should not exceed levels recommended by CODEX.
- If recycled water is used, it must be treated in a way that it does not become a risk to the safety of fresh pomegranate.
- Use only those waxes, fungicides at post-harvest stage, which are safe and approved by CODEX and should be in accordance with recommendations.
- Sprayers and automatic nozzles should be calibrated to control the accuracy of rate of application and properly cleaned to avoid contamination.

- The air cooling system be designed and maintained in a manner to prevent contamination of fresh pomegranates.
- The temperature of cold storage should be controlled and monitored to minimize microbial growth.
- The cooling system should be maintained in a clean and sanitary condition.
- Cleaning schedule of equipment and pack-house should be strictly implemented.
- All aspects of personnel hygiene in the pack-house should be strictly enforced.
- All hygienic aspects during transportation of fresh pomegranates in Reefer vans or containers must be looked into and no room for contamination needs to be left.

(Courtesy Post Harvest Manual for Export of Pomegranate, APEDA, 2002, New Delhi)

PROCESSED PRODUCTS OF POMEGRANATE

Sound pomegranate fruits fetch a fairly good price in the market and are hardly used for processing purpose. Hence, limited information is available on the utilization of pomegranate fruits for processed products (Elyatem and Kader, 1984). During seasonal glut, the prices of fruits are fairly low. Cracking of pomegranate fruits hampers their marketability. Such fruits can be utilized for processing. Pomegranate fruit with large juice grains can give an attractive coloured (purplish red) juice with mild acid sweet taste and low tannin content (less that 0.25%) are the desirable characteristics for processing of pomegranate.

Juice

Pomegranate juice makes a delicious drink (Gabunia et al., 1984). The juice may be extracted from whole fruits or from grains. On whole fruit basis the yield of juice is about 42 per cent, while from grains the yield is about 70 per cent (Siddappa, 1943). For extraction of juice, the fruits are cut into quarters and pressed under moderate pressure in a rack and cloth hydraulic press to recover 367-41 per cent juice (Saxena et al., 1987). The juice is also extracted from the seeds by pressing in a basket press. The juice obtained by pressing grains contains less tannins (0.12%) than that obtained from whole fruits (0.175%) Siddappa, 1943). The excess tannins in the juice are precipitated by gelatin. The juice is clarified by pectinase enzyme or heating in a flash pasteuzier at 79-82°C, cooling settling for 24th, racking and filtering. The clear juice can be preserved by treatement or by using chemical preservatives. The juice may be preserved by addition of 600 ppm sodium benzoate (Siddappa, 1943). The use of SO_2 or potassium metabisulphite is not recommended for pomegranate because of loss of its natural colour due to the bleaching action of SO_2. The chemical composition of pomegranate juice has been reported by several workers. The juice has been reported to contain 16 to 17 per cent total solids (Veres, 1976) and 0.81 to 1.23 per cent acidity as citric acid (Chavan et at., 1985). A technology for obtaining canned pomegranate beverage is described (Gabunia, 1984).

Wine

A commercial pomegranate wine made in California was reported to contain 11.9 per cent alcohol, 8.85 g/100 g total acidity (as citiric acid), 0.07g/100 ml volatile acidity and 11.5o Brix (Kainsa and Gupta, 1979). For preparation of wine the whole pomegranate fruits are pressed without crushing. Sugar is added to the juice to bring it to 22-23° Brix. Potassium metabisulphite is added to juice to prevent growth of undersirable micro-organisms. The juice is fermented with starter wine years. The wine is aged and finished in the same manner as in red pomegranate wine. Research carried out in our laboratories have shown that good quality wine can be prepared from pomegranate juice (Adsula et at, 1992). The rate of fermentation of pomegranate juice was, however, slower than that of grape juice.

Jelly and Anar-rub

An attractive jelly can be prepared from pomegranate juice (Saxena et al., 1979). A product known as anar rub with fairly good keeping quality can be made by concentrating pomegranate juice by adding sugar and heating the mixture on slow fire for long period. The finished product has thick consistency and contains 70-75 per cent total solids (Siddappa, 1943).

Anardana

Pomegranate seeds, particularly of sour types, can be dried and sold as anardana. It is used as an acidulant in place of tamarind or dried green mango for cooking purposes in North India in Indian style curries, chutney and other culinary preparations (Saxena et al., 1987). The anardana has high acidity (Pruthi and Saxena, 1984) ranging from 7.8-15.4 per cent Pruthi and Saxena (1984) recommended the use of cross -flow driers to get a uniform, hygienic, good quality anardana. The use of solar drier for dehydration of anardana has been reported (Saxena et al., 1987).

Rind Powder

This can be used as tooth powder and in medicine and cosmetic industries. A dry powder has been successfully prepared from pomegranate skin. Various steps in the preparation of pomegranate skin powder has been outlined in fig.3 . The recovery of powder was found to be 13.3 per cent (* haven, et al., 1985). This is an excellent source of B-carotene, potassium, phosphorus and calcium (Table). The powder contained 16.5 per cent polyphones and 5.35 per cent mineral matter.

Table: Chemical and mineral composition of pomegranate fruit and rind powder

Constituent	Fruit	Rind Powder
Moisture (%)	78.00	8.40
Protein (%)	1.60	0.94
Total sugars (%)	14.60	3.20
Ascorbic acid (mg/100g)	16.00	2.40
B-carotne (ppm)	-	12.52
Ash(%)	0.70	5.35
Polyphones (%)	-	16.50
Acidity(%)	0.58	4.13
Calcium (mg100g)	10.00	35.90
Phosphorus (mg/100g)	70.00	1100.00
Magnesium (mg/100g)	44.00	140.00
Potassium (mg/100g)	133.00	1550.00
Sodium (mg/100g)	0.90	40.00
Iron (mg/100g)	0.89	30.00
Zinc (mg/100g)	0.82	30.00
Manganese (mg/100g)	0.77	2.00
Copper (mg/100g)	0.34	1.00

FIGURE : 3 PREPARATION OF POMEGRANATE RIND POWDER

Selection of mature healthy pomegranate fruits

Washing

Separation of seeds and rind

Rind

(Courtesy Production Technology of Arid & Semi-Arid Fruits, MPKV, Rahuri, 1996)

POMEGRANATE PHYSIOLOGY

The fruits are ready for harvest between 135 and 170 days after an thesis. Fruits are harvested when the skin turns slightly yellow and the fruits give a metallic sound when tapped. At maturity the fruit colour changes in summer to dark yellow and in winter to dark red. The buds at the anterior end of the fruit curve inwards and become hard and dry at maturity. Properly matured fruits are easily scratched with a finger nail. The fruit is generally clipped from the tree when ripe and its quality improves on storming. Early harvesting in order to avoid cracking is one of the causes of poor quality of pomegranate in India.

Depending on the 'Bahar' treatments, pomegranate comes to harvest in installments. Usually the harvest commences in December/January and extends up to June/July. Crops ripening from April to June often get sunburnt and may also crack if rains intervene or irrigation is irregular. The pomegranate cv. 'Wonderful' fruit reaches horticultural maturity for commercial harvest when the total soluble solids (TSS) content attains a fairly constant level of 15%. The level of titratable acidity varied from one location to another and from one year to the next but generally remains stable at the time when the T.S.S. content reached 15%. After harvest, there was no further change in either TSS content or titratable acidity at 20°C, but redness of the juice continues to increase upto the harvest.

The development of pomegranate fruit under continental (Bet Shean Valley) and moderate maritime (coastal plain) climatic conditions. They observed that fertile flowers were vase-shaped and developed fruits, while bell-shaped flowers contained few egg cells and were sterile. Gibberellic acid induced the sterile flowers to develop into small fruits; however, these were devoid of juicy seeds. The growth of 'Mule's Head' pomegranate followed a single sigmoid curve, whereas in cv. 'Wonderful' the growth was more linear. The seeds accounted for about half of the fruit weight. The edible juicy tissue of the seeds grew continuously from June to October, whereas the internal stone tissue ceased growing and hardened by the end of June, Juice, TSS and anthocyain content of the fruit increased continuously during maturation, while acidity devised. Fruits of cv. 'Mule's head' ripen early and have a low acid content, whereas fruits of the late ripening cv. 'Wonderful' have high juice, TSS, acids and anthocyanins and are therefore suitable for processing. In both cultivars fruits colour developed gradually and served as a criterion for picking. The stage at which 70-90% of the skin is red corresponds with TSS: acid ration suitable for commercial picking. Some cultivars, such as 'Malissi', do not develop any red colour in the skin however.

In cv. 'P-23' (a seedling selection from the local Muscat') the initial elongated oval shape of immature fruit chānged at maturity to spherical, with marked depressions on the sides, and the green-purple rind colour turned to deep pink with reddish and yellowish pathos. The early milky-white colour of arils turned to a creamy pearl-white when the fruits matured. With the advancement of maturity, the percentage of arils increased while that of seed, rind and rind thickness decreased gradually. The TSS: acid ratio increased from 14.96 at 30 days to 64.83 at 165 days after nathesis.

From the pattern of CO_2 and ethylene production rates, pomegranate, fruits are judged to be non-climacteric fruits and thus maturation and ripening should take place on the plant before harvest to get quality fruits. The mature pomegranate may remain on the tree for a long period and picking time can be selected according to commercial needs. The harvesting of pomegranates in California begins in the middle of September and harvest maturity is judged on the basis of juice colour (the red colour must be equal to, or darker than, Munsell colour—chart 5 R-5/12), titratable acid content (<1.85%) of the fruit.

The red colour of pomegranate peel and juice is due to presence of anthocyanins. The most common anthocyanin in the juice is delphinidin-3, 5-glucoside and delphinidin-3-glucoside were also noted. The fruit rind contained large amounts of excessively astringent tannin, which made the juice unpalatable if the whole fruit is crushed or pressed at high pressure. The fruits 45-61% juice is calculated on a whole-fruit basis, or 76-85.5% in relation to the weight of arils. The juice has 12-16% sugar, consisting mainly of glucose or fructose; citric and malic acids have been identified as the predominant acids in fruits.

Post harvest storage

Pomegranate fruits are susceptible to moisture loss and need to be stored at high humidity. After harvest, the fruits are graded according to size, wrapped in paper and packed in bambook baskets or corrugated boxes. In bulk storage, fruits are packed in layers in wooden crates, each containing about 16-18 kg of fruits. Dry grass, rice straw or paper are used as cushioning material. In the state of Maharashtra (India) pomegranates are sorted into four grades, viz, super, king, queen and prince. The fruits are then packed in corrugated fiberboard (CFB) boxes. Four of five super-size frutis, six king size, nine queen-size or twelve prince-size fruits will generally fill a box. Innovative packaging developed at the Indian Agricultural Research Institute using ventilated CFB boxes with CFB partitions having vent holes and layer separators could be very useful for packaging pomegranates.

Pomegranate fruits have better keeping qualities than other tropical fruits such as mangoes, grapes and bananas. The storage life of pomegranates is comparable to that of apples. They can be stored for some months in a cool dry place, and upto six months under cold storage. The fruits, however, may be spoiled during transit or before marketing by rotting or the development of black colour in the fruits. The latter is thought to be due to post harvest biochemical changes.

Pomegranates can best be stored at low temperature and high humidity. Fruits stored at 4.5° C and 80-85% RH did not undergo any shrinkage or spoilage in a few months. It recommends an RH of 85-90% for storing 'Kandhar' pomegranates. Storage at lower temperature results in chilling injury, characterized by discolouration and pitting of the rind internal browing of the pith, paleness of the flesh and increased susceptibility to decay. Storage at 10°C is satisfactory if a post-harvest fungicide is used. Packing material containing sulphur compounds are found to give fruits 50% more protection from pathogens.

The control of RH is critical in the storage of pomegranate fruits. At low humidity, the skin desiccates readily and the rind becomes dark and hard; the fruits become less attractive and poor marketability. Storage at 5°C or lower resulted in chilling injury to the fruits. The severity of the symptoms increases with the exposure time and at temperatures below 10°C The peel of 'Wonderful' pomegranate fruit undergoes browning during storage below 10°C.

The storage life of Banling' pomegranate could be extended up to 12 weeks by keeping in sealed polythene bags at 10°C, with slight changes in quality such as weight loss, TSS, tritratable acidity. Individual shrink-warping of fruits and storing at 9° C gave 70 days of shelf-life without affecting the quality. Fruits dipped in 4% $CaCl_2$ solution for 3 minutes and stored at 2-4°C and 85-90% RH for 110 days showed low storage loss and retained quality. Fruits dried at 10°C RH, or

10°C and 47% RH, remained acceptable for three months or more during storage depending on the cultivar. Drying affects the quality but partially dried pomegranate may be useful for processed products.

Post harvest Physiology

The respiration rate of 'Wonderful' pomegranates remains low (less than 8 ml CO_2 $kg^{-1}h^{-1}$) when stored at 0-10°C for three months; it increases with temperature, however. The Q_{10} values for respiration are 3.4 between 0 and 10°C, 3.0 between 10 and 10°C and 2.3 between 20 and 10°C The rate of ethylene production from fruits stored below 10°C is mostly below 0.1 up $kg^{-1}h^{-1}$. Chilling injury symptoms and increased susceptibility to decay organism become more apparent after transfer to 10°C, internal symptoms are manifested as pale colour of the arils. Fruits held at 5°C for eight weeks show only a slight brown discolouration of the oculars speta. The temperature during storage for up to three months has little effect on soluble solid content, pH and titratable acidity of the juice.

Physiological Disorders

When the physiological disorder known as 'internal breakdown' occurs in the pomegranate, the pulp-bearing seeds (arils) do not develop the typical red colour and are somewhat flattened rather that plump. Flavour of the arils is abnormal and many have a streaked appearance due to fine white lines radiating from the seeds. There are no external symptoms. It originates during growth in some seasons, usually only in limited areas.

The incidence of internal breakdown develops 150 days after anthesis in variety G-137, and its intensity increases in the fruits which are left on the tree upto 165 days. The incidence of browning increases with increase in weight of fruit from 150 to 200 g (26.60 %) to more than 350 g (60%)

TSS, acidity, ascorbic acis, total sugars, reducing sugars, calcium, phosphorus and the enzyme catalyses were low whereas non-reducing sugars, starch, tannins, nitrogen, potassium, magnesium, boron, polyphenoloxidase and peroxides enzymes were high in affected arils of cvs. 'Ganesh' and 'P-23' than in healthy ones.

A superficial browning disorder (scald) develops on the husk of 'Wonderful' pomegranate fruit during storage. The severity of this disorder can be diminished by delaying the harvest time and by reducing storage temperature but these measures were insufficiently effective for storage periods exceeding six weeks and at temperatures of 9°C or lower chilling injury also occurs. Scald incidence was correlated with the amount of 0-dihydroxyphenols extractable from the husk, and was significantly controlled by measures that inhibited their oxidation by poly phenoloxidase. Such post harvest measures included dipping the fruit in boiling water for 2 minutes and in anti-oxidant solutions in biodithiocarbamate for 30 seconds, or by storing the fruit in a low-oxygen atmosphere. The most effective control of husk scald was obtained by storing late-harvested fruit in 2% O_2 at 2°C, but this treatment resulted in accumulation of ethanol, which caused off-flavors. When the fruits were transferred to air at 10°C both ethanol and off-flavours dissipated.

Post harvest Pathology

Three major diseases of pomegranate caused by micro-organisms are grey mould rot, heart rot and penicillium rot. Grey-mould rot is caused by Botrytis cinerea. Decay usually starts at the calyx. As the disease progresses, the skin becomes light brown, tough and leathery. The pulp-bearing seeds disintegrate into a dark mass in advanced infections. Under moist conditions, a characteristic grey mycelium appears on the affected surface.

Heart rot is caused by Aspergillus niger and Alter aria sp. Affected fruits show a slightly abnormal skin colour but internally a mass of blackened arils becomes prominent. Usually there is a black line of decay extending from the calyx to the interior part of the fruit. The disease develops while the fruit is on the tree. Affected fruits can usually be detected by serters and eliminated from the commercial pack. The rot due to Alternaria salani caused damage to fruits during storage and transit.

Penicillium rot, caused by P. expansum and other Penicillium spp., produces watery areas at the infection site followed by masses of blue or green spores. Infections in variably occur at skin breaks caused by cracking, mechanical injuries or insect punctures. Other fungus infect the same injured area and eventually overgrow the Pencillium. Other organisms causing decay in pomegranate fruit includes species of Botrytis, Cladosporium, Phoma, Phomopsis, Rhizopus and Sphaceloma punicae.

Dipping treatments with aqueous Topsin-M0. (1%) and Bavistin (0.05 to 0.1%) inhabited the growth of Aspergillus niger.

Infection of pomegranates with P. reticulosum causes the inner parts to develop grayish-green spore masses without any visible external symptoms. The possible toxicity of metabolic products of this fungus were investigated using Paramecium caudatum and white mice: the fungal metabolites caused the death of P. caudatum within 3-20 min. and of mice within 3-7 days after oral administration. P. reticulosum Birkinshaw has been known previously as a producer of mycotoxins. Further attention should be paid to its occurrence and possible hazards of its contamination of fruit, particularly that for juice production.

In view of the great importance of the pomegranate and considering its immense potential for cultivation under various adverse soil and agro-climatic conditions, more research efforts are needed for its inclusion in land-use management system, particularly on arid and semi-air lands. Apart from their well-established medicinal and curative properties, pomegranate fruits are also a rich source of minerals (calcium, magnesium and phosphorus). There are many wild forms of pomegranate grown in the sub-Himalayan areas which need proper exploitation for utilization into value-added products. Maturity indices based on physical criteria need to be established with respect to each variety. Pomegranates, being a non-climacteric fruit, have a tremendous potential for modified-atmosphere packaging (MAP) using various polymeric films which will not only retain fruit quality during storage but will also help in alleviating of chilling injury during refrigerated transport and storage. Therefore, an integrated approach on both production and post-harvest management using recent technologies on post-harvest handling, viz, individual shrink-wrapping, waxing, controlled-atmosphere (CA) storage coupled with judicious temperature management practices needs more attention for wide distribution of this delicious fruit in the global market.

(Courtesy S.K. Roy and D.P. Waskar MPKV, Rahuri)

PLANT GROWTH SUBSTANCES

PLANT HORMONES: STATUS AND DEFINITIONS

Plant hormone has been defined as organic substance produced naturally in the higher plants, controlling growth or other physiological functions at a site remote from its place of production and active in minute amounts. To distinguish it from animal hormone. It was termed as phytohormone. However, has started a discussion about abolishing the name of plant growth substances as hormone. This idea was hotly debated at the XI International Ceonference of Plant Growth Substances. The crisis of the phytohormone theory is caused mainly by the following problem: a poor correlation was found between growth rate and endogenous hormone levels in developing plant. Further, unexpected results were frequently found with exogenously applied hormone, i.e., in some cases, promoters caused an inhibition that inhibited a promotion of growth. Plant growth regulators are defined as organic compounds other than nutrient which in small amount promotes/inhibit or otherwise modify any physiological response in plants. The definitions adopted by the American Society of plant Physiology in 1954.are as follows:

Plant hormones are regulators produced by plants which in low concentration regulate a physiological plant processes.

Hormones usually move within plant from site of production to site of action. There are five classes of phytohormones: auxins, gibberellins cytokinins, abscistic acid and ethylene.

Auxins: These are organic substances which at low concentration (less than 0.001 M) promote growth along the longitudinal axis, when applied to shoot to plants freed as far as placing from their own inherent growth promoting substances.

Gibberellins: These substances are having gibbane ring skeleton capable of producing the same physiological responses as gibberellic acid that it must be active in specific gibberellin bio-assay. The gibberellins are phytohormones which are active in regulating dormancy, flowering, fruit setting and stimulating germination of seeds and extending growth of shoots.

Cytokin These are substances composed of hydrophilic group of higher specificity (adenine) and one lipophilic group of plant hormones having similar effects as those of GA in breaking the dormancy of a wide range of seeds and increases fruit set. These hormones mainly stimulate cell divisions and prevent chlorophyll degradation.

Various synthetic and natural plant growth regulators are listed below.

Names and structures of plant growth regulators.

Abbreviation or Common Name	Chemical Name or Structure	Major uses
ABA	Abscisic acid	Defoliant
ACC	1-aminocyclopropane-1-carboxylic acid	
Acifluorfen	5-2-Chloro-4-(trifluoromethly) phynexly-3-nitrobenzoic acid	GR Herb.
AVG	Amioethyoxy vinyl glycine.	
Alachlor	2-chloro-2', 6'-diethyl-N-(methoxymethyl) acetanilide	Herb.
Ancymidol	a-cyclopropyl-4-methoxy-a-(pyrimidin-5-yl) benzyl alcohol	GR
Atazine	2-chloro-4 (ethylamino)-6-(isopropylamino)-1,3,5-triazine	
BSAA	(Benzo-b-selenienyl)-3-acetic acid	DB
Barban Brassinolide	4-chloro-2-ynyl-3-chlorocarbanilalte	Herb.
Chlormequat, CCC	2-Chloroethyl-trimethyl ammonium chloride	LP,SR
Cycloheximide	3-{2-(3,5-dimethyl-2-oxocyclohexyl)-2 hydroxyethyl}-glutarimide	AS
3-CPA	2-(3-chlorophenoxy) propionic acid	FT
4-CPA	4-chlorophenoxy-acetic acid	FT
CPPU	N-(2-chloro-4-pyridyl)-N'-phenylurea	Herb.

Abbreviation or Common Name	Chemical Name or Structure	Major uses
2,4-D	(2,4-dichlorophenoxy) acetic acid	Herbs
Daminozide SADH	Succinic acid-2,2dimethyl hydrazine	GR
Dichlopentez ol	(E) - 1(2,4-dichlorophenyl)-4-4-dimethyl-2(1,24-triazol-1-yl)-1-penten-3-ol	GR Fung
Dikegulac	Sodium-2,3-4,6-di,I-isopropylidene-α-L-xylo-2-hexulofurasonate	GR
Diquat	6,7-dihydrodiprido pyrazinediium dibromide	FS
Diuron (DCMU)	3-(3,4-dichlorophyeyl)1,1-dimethylurea	
DPX-1840	2-(4-methoxphenyl)-3, 3a-hydro-8H-pyraxolo {5,1a} isoindol-8-one	GR
Etacelasil	3-chloroethyl-tris-(2-methoxyethoxy)silan	FL
Ethephon	2-chloroethylphosphonic acid	SR
Glyphosate	N-N-bis(phosphonomethyl) glycine	SR Herb
Glyphosine	N,N-bix (phosphonomethyl)glycine	SR
GA_3	Gibberellic acid	GE
Kinetin	6-Furfury laminopurine	-DB
IAA	Indole-3-acetic acid	Enlarge
IBA	Indole-3-butyric acid	Enlarge
Maleic hydrazide	1,2-dihydro-3, 6-pyridazinedione	Herb, GR

Abbreviation or Common Name	Chemical Name or Structure	Major uses
Mefluidide	N-2,4-dimethyl-5- (trifluoro methyl) sulfonylamino-phenyl acetamide	GR Herb
Mepiquat chloride	N,B-dimethyl piperidiunium chloride	GR
Metolachlor	2-chloro-N-(2-ethyl-6-methylphenyl)N-(2-methoxyl-1-methylethyl) acetamide	Herb
Metribuzin	4-amino-6-tert-butyl-4,5-dihydro-3-(methylthio)-12,4-triazine-5-one	Herbs
Monuron	3-(p-chloro-phenyl)-1,1-dimethylurea	Herbs
NAA	1-naphthalene acetic acid	FT GR
NOXA BNOA	2-nathpthhalenyloxyacetic acid	GS
Paclobutrazol	(2RS, 3RS) -1-(4-chlorophenyl)-4, 4-dimethyl-2-(1,2,4-triazol-1-yl)-pentan-3-olk	Gr
Paraquat	1,1-dimethyl-4,4-bipyridinium dichloride	Herb
Pretilachlor	2-chloro-2', 6'-diethyl-N-(2-propoxyethyl)acetanilide	Herb.
Propachlor	2-chloro-N-isopropylacetanilide	
Propazine	2-chloro-4-,-6,bix(isopropyl-amino)-1,3,5-triazine	
Release	5-chloro-3-methyl-4-nitro-1H-pyrazole	As

Abbreviation or Common Name	Chemical Name or Structure	Major uses
Terbutryn	2-(tert-butylamino)-4-(ethylamino-6(methylthio)-5 triazine p-hydroxyphenyl)-acetic acid	
2,4,5-T	(2,4,5-trichlorophenoxy) acetic acid	Herb
Tetcyclacis	5,(-4chlorophenyl) -3,4,5,9,10-pentaaza-terna-cyclo(5,410^{2,6}, O^{8,11}) dodeca-3,9diene	
Thidiazuron	N-phenyl-N'-1,2,3-thiadiazol-5-ylurea	CD GR
TIBA	2,3,5-triiodo-benezoic acid	GR
TRIA	1-hydroxy-triacontane triacontanol	GS
Triadimefon	1-(chlorophenoxy)-3,3-diamethyl-1-(1,2,4-triazol-1yl)-2 -butanone	GR
Triadimenol	3-(4-chlorophenoxy)-(1,1-dimethylethyl)-1H-1,2,4-triazole1-ethanol	GR Fiung
Triapenthenol	(E) (RS) 1-cyclohexyl 4,4-dimethyl 2 -(1 H 1,2, 4-triazol-1-yl) pent -1-en-3-ol	
Uniconazol	(E)0(4-chlorophenyl)-4-dimethyl-2-(1,2,4-triazol-1-ly)-1-penten -3-ol	GR Fung.

Abbreviations: AD= Apical-dominance reducer; AS=Absission stimulant: CD =Cotton defoliant; DB = Bormany breaker; Enlarg=Plant cell enlarger; FL = Fruit loosener; FS = Flowering suppressant, in sugarcane; FT=Fruit thinner; Fung = Fngicide; Gamet = Gametocide; GF = Grape enlarge; GR = Growth retardant; GS = Growth stimulant; Herb = Herbicide; LP = Lodging prevent or; PF = Pineapple floweing agent; SR = Sugarcane ripener.

Abscisic Acid (ABA): ABA is a naturally occurring sesquiterpene which regulate plant growth and metabolism in various ways and have been detected in nearly all plants. It is involved in the abscission of plant organs, induction and vegetative buds, in regulation of fruits' repeining and generally in reduction of growth.

Ethylene: It is the only gaseous hydrocarbon hormone which plays an important role in the ripening of fruits, inhibition of root growth, abscission and other growth processes. Unlike the other harmonies, ABA and ethylene are not discovered through any interaction with fungi.

List of hormones and their physiological responses

Effects	Auxin	GA	CK	ABA	Ethylene
Cell elongation	+	+	+	-	-
Cell division	+		+		-
Apical dominance	+		-		
Leaf growth					
In breadth		+			
Chlorophyll formation		+			
Vein elongation	+				
Leaflet movement	+				
Root formation	+				
Fiber growth in cotton seed	+	+	-		
Abscission	-	-		+	+
Sex expression					
Maleness	-	+		-	
Femaleness	+		+		
Fruit growth and division	+	+	+	+	+
Xylem formation	+	+			
Flower	+	+	+	-	-
Seed dormany		+	+	+	-
Bud dormany		+	+	+	-
Senescence	-	-	+	-	
Tropism	+	+		+	-

+Promote, -inhibit,

HOW HORMONE ACTS?

When a hormone acts upon a responsive plant system, it occurs, enters into some direct and molecular interactions, which results eventually in manifestation of measurable effects (bio-chemical or physiological responses). However, there are two aspects of hormone action.

1. **Mechanism of action:** The direct and specific molecular interaction, and
2. **Mode of action:** The succeeding series of steps (events and processes) which results in the measurable bio-chemical or physiological responses.

The two-term mechanisms and Mode of action should be taken as different terms rather than synonyms. The mechanism by which minute amounts of plant hormones, relatively simple organic compounds, can do dramatically controlled growth is one of the pressing and challenging problems in plant physiology. The progress in this field has been quite spectacular but still it needs a proper synthesis and analysis of various concepts given in this regards.

In the previous years numerous physiological and bio-chemical responses to plant hormones have been observed which have a bearing on the mechanism of hormone action. Some of the responses have been enumerated they are as follows:

1. Increased plasticity of the shoot and elasticity of the root cell wall.
2. Increased permeability of water.
3. The enhanced capacity to retain water taken up.
4. The active uptake of water and solute apparently independent of osmotic forces and occurring even against an osmotic gradient.
5. A decrease in protoplasmic viscosity.
6. An accelerated rate of respiration and cyclosis.

7. More rapid synthesis of protein, with lowered levels of free amino acids.
8. The increase of monosaccharide at the expense of reserve polysaccharide.
9 An increment of (intermicellar) wall pectin and of cellulose and a stimulation of alcohol and malic dehydrogenises, catalane, phosphates and ascorbic acid oxides.
10. These large varieties of effects which hormone evokes depict the multiplicity of sites.

An understanding of hormone action is dependent on location of sites of action. We are beginning to learn that the site of action of hormones is close to gene. Usually two sites of plant hormone action at the sub-cellular/molecular level are discussed; an early (primary) effect at the cell membrane and a subsequent (secondary) effect at the genetic level.

BIO-AGENT IN INTEGRATED PEST MANAGEMENT

Integrated pest control is an ecologically based pest population managment systems which uses all suitable techniques to reduce or to manipulate the pest population in that it prevents from causing economically unacceptable injury to crop. Use of bio-agents such as, predtators, parasites and pathogen along with the combination of chemical, physical, mechanical control of pest is a must to bring down the population and maintain balance of nature.

Some of the Promising Bio-agents

Trichoderma: Trichogrammatids are one of the most important group of biotic agents. In India about 26 trichogrammatids are recorded. Trichlogrammatids have a wide host range. In case of Pomegranate butterfly it gives 33.3-93.3 per cent control with parasitism's in June-July.

Green lace wing: (*Chrysoperia carnea*)

Green lace wing is being used in control of Aphids, white fly meals bugs.

Graytoleamus montruniers: The mass production and period of release against mealy bugs of fruits crops has been going on for 60 years. They are reared on red pumpkin in laboratory, like cycle of adult takes after 30 days. Colonies of 30-40 adults coccinellids on each pumpkin, when mealy bugs are 8 days old. They deposit their eggs on pumpkin. Emerging beetles are collected. Approximately 21 days adult predators, about 20 per tree, are released and each tree receives 1-5 releases annually.

Pests	**Biological agents**
1 Pomeganate butterfly (Fruit borer) Vivacola isocrates Fab	• Hymnopterous parasols (a) Telenomus Sps (b) oocncyrtus Sps
2. Bark eating caterpillar Inderbella SPP	• Pathogen *Beauveria bassiana*
3 Mealy bugs *Ferrisla virigata*	• Predators
Psudoccousi lacines	Cryptolaemus Mountrozourei
4 Aphids, white fly, thrips	• Crysoperla carnea
5 Scale insect Paris sseita Niger	• Parasite Seutellisa cyanea
other fruit sucking moth otherois	• Parasite Telenomus Spp.

FERTIGATION OF POMEGRANATE:

Days	Grade	Quantity
1-15	12:61:0	1 kg day/AC
16-30	11:42:11	1 kg day/AC
31-60	19:19:19	1 kg day/AC
6	Urea	1 kg day/AC

Days	Grade	Quantity
61-90	15:15:30	1 kg day/AC
	Urea	1 kg day/AC
91:120	0:52:34	1.5 kg day/AC
	Urea	1 kg AC/day
	Next day Calcium Nitrate	1 kg AC/day
121-150	0:0:50	2 kg AC/day
	Next day Calcium Nitrate	1 kg AC/day

Pests	Biological agents:
1 Pomegranate butter fly (Fruit borer) Vivacola isocrates Fab	• Hymnopterous parasols (a) Telenomus Sps (b) ooencyrtus Sps
2. Bark eating Caterpillar Inderbella SPP	• Pathogen Beauveria Bassiana (Metabelidae lepi)
3 Mealy bugs *Ferrisla virigata*	• Predators
Psudoccousi lacines	Cryptolaemus Mountrozourei
4 Aphids, white fly, thrips	• Crysoperla carnea
5 Seale insect Paris sseita Niger other fruit sucking moth otherois	• Parasite Seutellisa cyanea i) Parasite Telenomus Spp.

Pomegranate Wine: Scope and Importance

Introduction

The value added products like pomegranate juice, jelly, anardhna, anarrub, rind powder etc., are prepared from pomegranate fruits. The 1000 g of pomegranate wine can be prepared through fermentation technology by using suitable wine yeast, (Ezyhrinciew, 1996; Probaglav et al, 1983; Shivanna 1997; singhnagi and Manirekar 1975; Suresh and Ethiraj 1968 Suresh et al. 1985). Wine is a fermented beverage from fresh fruits composed of water, alcohol, pigments esters, vitamins, carbohydrates, minerals, acids, tannins with medicinal and theropeutic value.

Importance of Wine:

Production of fruit based wines are helpful in the following ways.

- Helps to avoid post-harvest losses.
- Helps to get additional income to the growers.
- Reduces wastage of fruits.
- Provides healthy drink for relief and relaxation.
- Creals export market and earns foreign exchange
- Industrializes the region.
- Improves the economic condition of the people.
- Generates employment opportunities.
- Encourages to shift from hard liquor habits.
- Utilization undersized and spotted fruits.

There is a scope to popularize Indian diet with fruit wines both in India and abroad. Lower death rates from heart diseases are often found among wine drinkers in western world. Just as a good masala cannot be prepared with a single spice, the best wines and vermouth could be made from a blend of different fruits or different varieties of fruits. In European countries homemade wines are costlier than wines manufactured in wineries. The fruitgrowers can make fruit wines at home without much additional expenditure. World's popular wine brands are known by their regions of their origin. If developed, one day in the near future Indian wines could become globally popular.

Anti-transpirant

It is a material which is applied to transplanting plant surface with an aim to reducing the water loss from the plant (mainly leaf)

Approximately 99 per cent of water taken by the plant through roots is transpired to the atmosphere through the stomatal spores of leaves. The stomata bearing leaf surfaces are most important sites for the anti-transpirants applications. The types of anti-transpirants are described below:

Film forming types

Plastic film farming anti-transpirants are more permeable to CO_2 than H_2O under water stress e.g. higher alcohol, ethyl achohol, hexodecasol S-789 Tag-9)

Metabolic inhibitors

Decreases more transpiration (50%) but decreases the photosynthesis to a lesser extent by 35 per cent by closing stomata under moisture stress. E.g. phenyl methyl acetate (PMA) (fungicide) are most widely used as anti-transpirants.

Reflectant types

Reflecting coating raises the light saturation level and increases light compensation points of all species. Improves water use efficiency with increased radiation e.g. 2 per cent Koolinite spray.

Plant growth regulators

These are plant growth regulators which close the stomata under moisture stress e.g. ABA (abscissic acid) and 2,4,5-T.

METHODS OF IRRIGATION APPLICATION

i) **Surface irrigation**: Water is applied by complete wetting land surface as in border strip or check method or wetting only a part of the surface as in furrows or ring.

ii) **Sprinkler or overhead irrigation**: The soil is wetted in a similar fashion as by rainfall.

iii) **Sub-irrigation**: Water is applied beneath the land surface from shallow water table wetting the surface little or not at all e.g. perforated pipe below soil to maintain water table.

Water management studies in important fruit crops

Sl. No.	Crops	Irrigation scheduling approach	Layout	No. of irrigations	Water requirement (cm)	Remarks
1	Pomegranate	50% soil moisture depletion.	Check basin	30	200-225	Uneven interval of irrigation during latter period of fruit, depletion development and maturity induces cracking of fruits. Interval of 7-10 days of Mrig bahar and 5-7 days during Ambe bahar crop is recommended.

MATURITY INDICES

For getting quality fruits, they are harvested at proper stage of maturity. In Maharashtra, the maturity of pomegranate fruit is judged by the following indics:

1. The fruit requires 130-150 days after fruit set for proper maturity,

2 In summer the fruit colour changes from yellowish to dark red at maturity,

3 The fruit is harvested when it gives cracking sound when tapped.

4. The calyx at the distal end of the fruit gets closed at the time of maturity and

5 Properly matured the fruit is easily scratched with finger nails.

Effect of method of irrigation on growth and yield of pomegranate

Methods	Litres water used (5 Mar.-14 June)	Tree height (cm)	Stem dia (cm)	Fruit yield (q/ha)	W.U.E. (q/ha/cm)
Heck basin					
0.6 IW/CPE	48900	2.35	3.78	40.6	0.82
0.8 IW/CPE	6600	2.56	4.02	47.5	0.69
Brickle					
Daily, 20 %wetted area	3580	2.16	3.79	52.5	1.25
Daily, 30% wetted area	5322	2.44	4.02	44.9	0.72
Alternate day, 20% wetted area	359	2.30	3.84	48.6	1.16
Alternate day, 30% wetted area	5322	2.24	4.01	57.3	0.92

Performance of drip method of irrigation on different fruit crops

Sl. No.	Crops	Yield (q/ha)		Water applied (cm)		Water saving (%)	Increase in yield (%)
		Surface	Drip	Surface	Drip		
1	Pome-granate	53.5	69.7	218	121	45	12

Average water requirement of fruit tree by drip method (litre/tree/day)

Fruit Crops (spacing)	Age (yrs)	Water requirement (litres/tree/day)											
		Jan.	Feb.	Mar.	Apr.	May	June	July	Aug	Sept.	Oct	Nov.	Dec.
Pomegranate	1	-	-	-	-	-	-	13	10	10	12	9	8
	2	12	17	25	34	36	25	18	15	15	18	13	11
(5x5m^2)	3	16	23	34	46	50	34	25	21	21	24	19	16

CASE STUDY FOR POMEGRANATE ORCHARD

Design drip system of irrigation for 1 ha of an orchard planted with pomegranate (G-137). The spacing for pomegranate is 5 x 5 m. The available data is given below:

1. Size of land : 100 x 100 m
2. Type of soil : Light textured
3. Land slope : 0.30 per cent (S-N)
4. Maximum evaporation : 12.5 mm/day
5. Water source : Well
6. Available discharge : 3.5 lps
7. Static head : 10 m
8. Wetted area : 20 per cent

9. Age of orchard : 3 years

Net depth of water

i.e. Evapo-transpiration of crop (ET)

$E_T = E_p \times K_p \times K_c$

Where,

E_p = Pan evaporation (mm)

K_p = Pan factor (taken as 0.7)

K_c = Crop factor (depends on the age of the crop)

ET = 12.5 x 0.7 x 0.75

= 6.56 mm/day

The crop factors K_c and wetted area for different fruit crops are given as follows.

Sl No.	Fruit crops	First year		Second year		Third year		Fourth year		Fifth year on Wards	
		KC	Wa	KC	Wa	KC	Wa	KC	Wa	KC	Wa
1	Pomegranate (5 x 5 m^2)	0.40	0.30	0.50	0.35	0.60-0.65 (Full grown)	0.40	-	-	-	-

Volume of Water

Total volume of water required (mm) – Depth of water x Plant spacing (m) x Row spacing (m) x Wetted area (in fraction)

= 6.56 x 5 x 0.20

= 32.81 /day/ plant

Considering 90 per cent emission uniformity of the drip system, volume of water required = 36.51/ day/plant. Per plant wetted area is the area which is shaded due to its canopy cover, when the sun is overhead, which depends on the stage of the crop growth.

Emitter selection and positioning:

The age of pomegranate tree is 3 years i.e. the trees are fully grown. The soil is light textured.

Use 4 drippers, each of 4 lph dischanrge i.e. the total discharge rate = 16 1/h

Operation time of system

$$\text{Operation time} = \frac{\text{Volume of the water to be applied per tree}}{\text{Total dripper discharge (4 x 4)}}$$

$$= \frac{36.5}{16}$$

= 2.28 h = 2 h 17 min.

Drip Irrigation

The amount and distribution of rainfall in many parts of the world are inadequate to meet the total water requirements of crops. The only solution to this is by supplementing water through irrigation. Being a limited resource, efficient use of water through scientific irrigation management is of utmost importance in providing the best insurance against weather induced fluctuation in food production.

Drip or Trickle irrigation is one of the latest innovative irrigation methods. It is the slow application of water on or beneath the discrete or continuous drops, tiny streams or miniature spray through emitters or applications placed along a water deliver line near the plants. In this system water is delivered to each plant at it's root zone through a network of tubing.

PRINCIPLES OF DRIP IRRIGATION

Water is required only for the root zone of any crop and this can be achieved through drip irrigation system. It is a method of watering plants frequently and with a volume of water aapproaching consumptive use of the plants, thereby minimizing such conventional losses as deep percolation, run off and soil water evaporation. In this method irrigation is accomplished by using small diameter plastic lateral lines with devices called 'emitters' or 'drippers' at selected, spacing to deliver water to the soil near the base of the plants. The system applies water slowly to keep the soil moisture within the desired range for plant growth. Drip irrigation system facilitates in maintaining a constant moisture level at the crop root zone thereby making available a regular supply of water and plant nutrients throughout the growth making available a regular supply of water and plant nutrients throughout the growing plant of the crop. Drip irrigation method is characterized by the following features:

i) Water is applied at a low rate

ii) Water is applied over a long period of time

iii) Water is a applied at frequent intervals

iv) Water is applied near or into the plant root zone

v) What is applied by a low pressure delivery system (0.2 to 2 kg/cm^2). Since the area wanted by each emitter is a function of the soil hydraulic properties, one or more emission points per plant may be necessary.

ADVANTAGES OF DRIP IRRIGATION

The potential advantages of drip irrigaition can be summarized as follows:

- Increased beneficial use of available water by irrigating only a smaller portion of the soil volume, decreasing surface evaporation and controlling on deep percolation losses.
- Enhanced plant growth and yield by keeping soil water content fairly constant in the plant root zone.
- Retarded weed growth since only a fraction of soil is wetted.

- Improved application of fertlizers by factilitating frequent split application with the irrigation water or otherwise.
- Decreased energy requirements since the quantity of water pumped is lesser than other irrigation methods and also due to the low operating pressure to drip irrigation system.
- Reduced operational labour when compared to conventional irrigation methods.
- Suitable for irrigation of hilly terrain and problem soils: If you take Kerela into consideration where high value plantation crops like coconut, arecanut, rubber, cardamom etc. are cultivated on a large-scale mainly on lans with undulating topography, the feasibility of adopting conventional methods of surface irrigation in these areas is less. In addition, the frequent occurrence of erratic monsoons and the existence of a definite dry spell during December-May season with problems of water scarcity necessitates economy in water use in this region. Drip irrigation is an ideal technique under such conditions.

Organo-mineral pits-A technique for soil moisture conservation

For better moisture conservation as well as application of irrigation water at root zone depth, under drip irrigation, water is applied in pits taken around the plant. In the case of coconut where root zone extends laterally to about 2.0 m radius, four pits are dug, each one at a distance of 1.5 m., form the palm of four sides of the basin. The size of each pit will be 0.5 x 0.56 x 0.5 m. These pits are filled with alternate layers of coconut husk, dry cow dung dry leaves and finally soil on top. These are called organo-mineral pits. Irrigation water is applied through the emitters into these pits. In order to enable the distribution of water to the lower portion of the pit s a sub-soil distributor is used.

Sub-soil distributor. This is a PVC tube (Half inch internal dia) of 25-30 cm length with perforatios made all around. This tube is inserted into the organo-mineral pit and emitter is placed into it. This will enable better distribution of water to the lower portion of the pits and at the active root zone depth.

SYSTEM LAYOUT

Layout of the drip irrigation system involves simple pipe arrangements and the decision as to the number and spacing of laterals is governed by the nature of the crop. Each row of plants, whether trees, vegetables or others will usually require one lateral. In some horticultural crops, double row planting allows two crop rows to be served by the lateral. The in-row placement of emitters has the distinct advantage of minimum interference with other cultural operations. If adequate wetting is not achieved by one lateral line, emitters may be installed on a double line or a loop (some-times called a pig-tail) around the base of the tree.

FERTILISER APPLICATION UNDER DRIP IRRIGATION

Soluble fertilizers may be distributed through drip irrigation system, while those that are only partially soluble should not be applied through the drip system. With the commerical drip irrigation systems, available in the market, fertilizer injection tanks are provided, whereby the soluble fertilizers can be injected into the irrigation water. The partially soluble/insoluble fertilizers may be applied manually inside the organo-mineral pits where the drip emitters are placed.

BAHAR TREATMENT

Bahar treatment means giving rest to the plants holding of water. In other words it it artificial stress creation within the plant by no supply of water and nutrients.

C:N ratio is normally 28-30 parts of C to 1 part of N. So the ratio is 28:1.

It is known and analysed that during flowering stage the ratio comes to 14:1 which means that there is depletion of CHO and not accumulation. The lower CHO means relatively high N and it is this relatively higher N that gives rise to flowering.

The stress period depends upon the type of soil or heavy soil, more the stress on light soil, lesser the stress period. Thus nutrient concentration is maintained

BAHAR TREATMENT INCLUDES

1. Water with holding
2. Nutrition and fertilizer use.
3. Change of soil in the root zone.

As the stress during bahar treatment are being resorted to there is not need of addition of nutrients. The initial nutrition plays role in giving profuse flowering.

If the tree has more concentration of balanced nutrients more is the flowering.

In ring and basin systems of irrigation, every year we add lots of nutrients out of which the portion of soil fed by hair roots is again replaced by ring soil. The reasons behind that is fertilizers are added to the root zone, which are used by the plants and soils around that portion which get depleted. Thus while giving bahar treatment the soils of these portions are interchanged. Hence every year we add enriched soil to the feeding zone and depleted soil is put into the ring zone. This is how soil fertility can be maintained.

Pruning of unwanted roots is also done through bahar treatment.

Hormones are organic chemical messengers that are produced in one tissue and term located to another where they produce their specific effect, and regulate plant growth and development.

The term hormone comes from Greek hormone and means "to excite" this might give impression that all hormones of plants are stimulatory to growth. Many hormones inhibit growth. Many synthetic compounds mimic the effect of hormones. Hence the term plant growth substance or growth regulator is used for any hormone-like compound, natural or synthetic. The term hormone is restricted to natural products only.

Characteristics of Plant Hormones

i) Plant hormones are produced most abundantly in the actively growing parts of the plant body such as apical meristems of the shoot and root, young growing leaves, or developing seeds of fruits.

ii) They are trasnsported both through·the vascular tissues and through non-muscular tissues such as parenchyma. The transport is in any one particular direction, hence it is polar transport. The direction is generally basipetal, i.e., from the tip to the base of an organ. It is probably acropetal, i.e., from the base to the tip in some.

iii) Only minute amounts of the substance are needed to influence a physiological process. They act in less than one-thousandth molar and often in one-millionth molar range.

iv) Unlike animal hormones, it is difficult to separate the site of production from the site of action.

v) Plant hormones may be stimulatory or inhibitory depending on the concentrations, i.e., they can have a positive or promoting effect or can have a negative or inhibiting effect.

vi) The same hormone may produce different effects, i.e., promote some processes, inhibit some others and may not produce any effect on others.

vii) Different hormones interact in complex ways to control

a particular physiological process (growth, development or metabolism).

Kinds of Plant Hormones

Just as a machine needs both an accelerator and a brake for effective and precise control, and also the growth systems of plants need growth-promoting substances and growth-inhibiting substances. There are five main groups of plant growth substances—three of them (auxins, gibberellins and cytokinins) are growth promoters and two (abscisic acid and ethylene) are growth inhibitors.

Growth Promoters

The Auxins

The auxins were the first plant growth substances to be identified. Naturally occurring auxins are weak organic acids. The acidic group is bound to the end of a side chain attached to a ring compound. The most common naturally occurring auxin is identified as indole acetic acid (IAA).

Natural and Synthetic Auxins

i) Naturally Occurring Auxins: The most common naturally occurring auxin has been identified as indoleacetic acid (abbreviated as UAA). It is derived from the amino acid tryptophan. It is produced mainly in the shoot apical meristems, young leaves and seeds. A very small amount is also produced in the root tips.

ii) Synthetic Auxins: Many substances similar to auxins have been synthesized in the laboratory. These include two phenoxyacetic acids: 2,4-D (2,4-dichlorophenoxyacetic acid) and 2,4,5-T (2,4,5 trichlorophenoxyacetic acid); IBA (indole bytyric acid); NAA (naphthalene acetic acid) and two benzoic acids: 2,3,6-trichlorobenzoic acid (TCBA) and 2,3,5-triiodobenzoic acid (TIBA) are the others.

Effects of Auxins

Auxins influence a number of activities of plants. The following are some of them:

i) **Cell Division and Differentiation:** The vascular combine is the site of response to auxins. Cell division in the cambium is stimulated, so also the differentiation of the newly produced cells into xylem and phloem elements. If apical mersitem is destroyed the basipetal flow of auxin stops and vascular tissue differentiation is inhibited.

ii) **Cell elongation:** Auxins stimulate elongation of cells in the young internodes just below the apical meristem. Auxins possibly act by actively transporting protons across the cell membrane and thereby activating the enzymes that weaken the bonds holding the cellulose molecules of the cell wall.

iii) **Apical Dominance:** The growing apical meristem of a plant prevents the growth of lateral buds. This is called apical dominance. If the apical meristem is cut off, axillary buds soon grow to produce lateral branches. If an auxin is applied to the cut surface of the stem, the buds will continue to be dormant. If a potato tuber is treated with auxin, the lateral buds ("eyes") will remain dormant. Hence an auxin maintains dormancy. Once the buds begin to grow, auxins stimulate their growth.

iv) **Root Initiation:** Auxins stimulate the growth of adventitious roots. If cut ends of stems are dipped in dilute solution of auxin, roots develop at the cut ends. New plants may be propagated this way.

v) **Flower Initiation:** Application of low concentrations of 2,4,-D and NAA initiate flowering in pine apple.

vi) **Control of Abscission:** Abscission is a process by which leaves and fruits drop off from a tree by the formation of a weak layer of cells called the abscission layer in their stalk. Young leaves and fruits are firmly attached to the stem as long as they produce auxins. When the quantity of auxin produced becomes less, the abscission layer develops at the base of the petiole or the fruit stalk. Soon the petiole or the fruit stalk breaks at this point and it falls to the ground.

vii) **Prevention of Premature Withering of Flowers and Fruits:** The spraying of 23,4-D,IAA and IBA prevents premature (pre-harvest) withering of flowers and fruits of orchard plants (citrus, pear and apple).

viii) **Induction of Parthenocarypy:** A fruit develops from the ovary of a flower. Normally a fruit develops only after pollination and fertilization. When auxins are applied to immature flowers, the ovary is stimulated to develop into fruit even before pollination. Such fruits develop without pollination or fertilization and are called parthenocarpic fruits. The developing embryo or seed itself is the source of auxin for the development of the fruit. If the seeds are removed during development, the fruit stops growing.

ix) **Sex Expression:** Spraying of auxins causes cucurbit plants increase the number of female flowers.

x) **Effect on Tropisms:** A growth response to light is phototropism. If a growing plant is illuminated by light on one side, it bends towards light; the reason for this is the inactivation of auxins on the illuminated side by light (photoinactivation). This produces unequal concentrations of auxins, higher on the shaded side, and lower on the illuminated side. Differential growth occurs and the shoot bends.

Unequal concentrations of auxins and the inhibitory action of auxins on roots is given as the cause for positive geo-tropism (now called gravitropism) in roots.

xi) **As Herbicides:** the synthetic auxins 2,4-D and 2,4,5-T are selectively toxic to unwanted plants called weeds, growing among crop plants, cereals and lawn grass.

Hence they act as weed-killers.

The Gibberellins:

Effects of Gibberellins

i) **Elongation:** When gibberellins are applied to mutant dwarf plants which have lost their capacity to synthesize gibberellins, they grow quite tall and become indistinguishable from the normal tall plants. Application of gibberellins to normal plants may also produce elongation. Too much gibberellin causes the stem to become long and thin with very few branches and pale stems.

ii) **Bolting and Flowering:** Biennial plants such as cauliflower and cabbage form leaf-restates before they produce flowers. Flowers are produced under certain specific photoperiodic conditions. Application of gibberellins causes stem elongation (upto 4-5 meters) and flowering in these plants. This phenomenon is called bolting. In a normal plant bolting occurs only in the second year of growth. Gibberellins can induce bolting in the first year itself.

iii) **Sex Expression:** It has been observed that gibberellins influence the formation of male or staminate flowers in cucumbers and squashes.

iv) **Breaking Dormancy:** Gibberllins break the dormancy periods of seeds, stimulating them to germinate. In the seeds of the Graminease, there is specialized layer of cells called the aleurone layer just below the seed coat. This is rich in protein. Just before germination, when the seeds absorb water, gibberellin is produced which diffuses into the aleurone layer. In response to gibberellin the aleurone layer produces several enzymes which speed up digestion of reserve food, weakens the seed coat, thus making it easy for the root to emerge out.

v) **Parthenocarpy:** Gibberllins can induce the formation of parthenocarpic fruits and are more potent than auxins in this context.

vi) **Hybrid Vigour:** Gibberllins play a role in hybrid vigour. An analysis of gibberellins of hybrid and inbred strains of plants, in 1988 showed that hybrids produce higher levels of gibberllins. When additional gibberellins are applied to inbred plants, their growth also approaches that of hybrid plants.

Cytokinins produce the following effects:

i) **Cell Divison:** As has been established in various tissue culture experiments, the most important function is stimulating cell division. It also promotes cell elongation.

ii) **Organ Formation:** Cytokinins also plays a role in organ formation, i.e., morphogenesis of roots and shoots. Varying concentrations of cytokinin/auxin ratios can induce development of roots/shoots or both.

iii) **Lateral Bud Development:** When cytokinin is applied to a non-growing lateral bud, it is stimulated to grow.

iv) **Chloroplast Development and Chlorophyll Synthesis:** Application of cytokinins to etiolated leaves, enhances development of chloroplast, especially formation of grana and increases the rate of chlorophyll synthesis.

v) **Delay in Sensecence:** Along with auxins, cytokinins prevent premature ageing or senescence by stabilizing proteins and chlorophylls.

vi) **Breaking Seed Dormancy:** Cytokinins are known to break seed dormancy and induce germination.

Growth Inhibitors

Abscisis Acid:

The best known growth inhibitor is abscises acid (ABA). The name is derived from the ability of the substance to promote abscission, a discovery made by F.T. Addicott. It is synthesized in fruits and leaves and has important effects on growth and development.

i) **Abscission:** It promotes senescence of leaves. It may play a role in formation of abscission layer at the base of a leaf flower on fruit. The abscission layer breaks to cause the fall of leaves and fruits.

ii) **Dormancy:** It suppresses elongation of shoots, inducing the formation of dormant buds.

iii) **Stomatal Closure:** It functions as a stress hormone and helps to cope with drought by stimulating closure of stomata (see, section 4-14)

iv) **Defence against Salt and Cold Stress:** When plants are under stress in conditions of inadequate water supply, salinity and chilling and freezing temperatures. ABA level increases in plants, suggesting that it helps plants to overcome the 'stress' to some extent.

v) **Embryo Development in Seeds:** ABA inhibits development of embryos in seeds.

Ethylene

Ethylene is the only gaseous or volatile plant hormone. It is a product of metabolism of the amino acid methionine and is produced in nodes in stems, ripening fruits and senesceince tissues. It is released from most plant organs of a varying concentration, especially from ripening fruits.

i) **Fruit Ripening:** Ethylene is a fruit ripening agent. It is released in large quantities during the ripening phase called climacteric. When complex carbohydrates are broken down into simple sugars, chlorophyll breaks down, cell walls become soft and volatile aromatic substances are released.

ii) **Triple Response:** It inhibits elongation of stems, increases stem thickening and promotes horizontal growth habit.

iii) **Senescence:** It promotes senescence of flowers and leaves and their abscission.

iv) **Additional Effects:** It initiates formation of root hairs, latex production and the formation of aerenchyma in submerged root and stems.

Plant Hormones and some Known Effects

SI. No.	Hormone	Effects
1	Auxin	Promotes cell elongation; formation of roots in cutting; apical dominance; fruits and muturation; xylem differentiation; tropism
2	Gibberellin	Promotes stem elongation (in dwarfs); involved in flowering; release of dormancy of buds and seeds; stimulates growth of pollen tube during fertilization.
3	Cytokinin	Cell division; bud embryo development; slows down leaf aging.
4	Abscisic acid	Promotes stomatal closure; initiates bud dormancy; probably not invovled in abscission
5.	Ethylene	Promotes fruit ripening; promotes abscission of leaves, flowers and fruits; initiates root hair, latex precaution and formation of aerenchyma in submerged roots and stems.

Application of Phytohormores in Agriculture and Horitculture

Plant growth substances especially synthetic auxins, and to some extent, ethylene have found immense uses in agriculture and horticulture. Some are mentioned below:

i) **Rooting of Cutting:** When cuttings of woody plants are to be planted, it is essential that they sprout adventitious roots when planted. If the cut end is dipped in a solution of a synthetic auxin like NAA (naphthalene acetic acid,) roots arise quite quickly. A successful development of new plants can thus be achieved.

ii) **Fruits-setting** NAA and IbA have been extensively used to improve natural fruit-setting in trees in orchards.

iii) **Flower Initition:** NAA, 2,4-D (s,4-dichlorophenoxy acetic acid), when sprayed on leaves (foliar sprays) in pineapple, apple, and pears helps to cause flowering and also fruit-ripening.

iv) **Dormancy in Potato Tubers:** To prevent the lateral buds ("eyes") of potatoes from sprouting during storage, agriculturists use sprays of IBA and methyl esters of NAA. With this treatment potatoes can be stored for as long as three years.

v) **Premature Fruit Fall;** 2,4-D can counter the effects of abscisic acid (that promotes the dropping of fruit from trees). Application of a spray to pear and apple trees is a practice to make the trees retain and ripen more of their fruits.

vi) **Weed-Killing:** 2,4-D; 2, 4, 5-T; 2, 3, 6-trichlorobenozoic acid (TCBA) and 2, 3 5, -triiodobenzoic acid (TIBA) are used as selective weed-killers. Most weeds are fast growing dicotyledons; they are undesrirable. In a field of monocot crops, if the weed-killers are sprayed in certain dosage, the weeds grow abnormally, grossly deformed, and finally die. Monocots are not visibly, affected. 2,4,5-T was particularly effective against woody perennials of rough pastures. It has been banned since it contains traces of toxic dioxin, which can cause foetal abnormalities, a severe form of acne (chlorachne) and cancer.

A mixture of 2,4-D and 2,4,5-T was sprayed from aeroplane is Vietnam War by USA to defoliate (-to remove leaves) vast areas of countryside. The contamination was detected in drinking water and fish. It was also the reason for abnormal development of some babies. Later, this programme, called 'Agent Orange', was given up.

vii) **Fruit Ripening:** Ethylene gas is released slowly from a chemical called ethaphon, (commercially available as Ethrel). It is 2-chlorethyl phosphoric acid which rapidly breaks down in water at alkaline pH to release ethane. It stimulates the production of enzymes that promote fruit-ripening. If fruits in tropics are to to be sent to far off places, fruit growers can pick citrus fruits, bananas and pineapple when they are still green. They can be ripened by application of ethane after they reach their destination.

Table 1 : Production cost of homemade Pomegranate Wine:

Sl. No.	Items of wine making	Cost/quantity
1	Cost of 1 kg fruits (Rs)	10.00
2	Yield of juice (ml)/kg fruit	500.00
3	Cost of added chemical (Rs)	2.00
4	Processing cost (1/3rd of raw material cost, (Rs))	4.00
5	Wine yield from 1 kg fruit (ml)	400.00
6	Cost of one litre wine (Rs)	40.00

Table 2: Organoleptic Evaluation of homemade Pomegranate Wine

Sl. No.	Quality parameters	Dry wine	Sweet wine
1	Appearance (2)	2.00	2.00
2	Colour (3)	2.50	2.50
3	Aroma (2)	2.00	2.00
4.	Bouquet (1)	1.00	1.00
5	Vinegar (2)	1.50	1.75
6	Total acidity (2)	1.50	1.75

Sl. No.	Quality parameters	Dry wine	Sweet wine
7	Sweetness (1)	0.00	0.50
8	Body (1)	0.50	1.00
9	Flavour (2)	1.00	1.50
10	Astringency (2)	1.00	1.00
11	General quality (2)	2.00	1.00
12	Overall acceptability (20)	2.00	1.00

(Figures in the paranthesis are maximum allotted marks for each parameters).

Table 3: Organoleptic evaluation of pomegranate sweet wines (mean average of three judges)

Wine	Spices added	Scores given by panel of judges in 20 point scale.											
		Appearance (2)	Colour (3)	Aroma (2)	Boaquet (1)	Vinegar (2)	Total acidity (2)	Sweetness (1)	Body (1)	Flavour (2)	Astringency (2)	General quality (2)	Overall acceptability (20 points)
Ganesha	Clove	0.50	1.00	1.00	1.00	2.00	1.00	1.00	0.50	2.00	2.00	1.50	12.00
	Cardamon	1.00	2.00	1.50	1.00	2.00	2.00	1.00	1.00	1.00	1.00	2.00	15.50
	Ginger	1.00	1.00	1.00	1.00	0.00	1.00	1.00	1.00	2.00	2.00	2.00	14.00

Fig-1: Flow chart of wine preparation from Pomegranate:

Pomegranate Fruits

Washing

Aril Separation

Squizing of arils

Juice Extraction

Addition of Potassium metabisulphite Keep for 4 hours and raise the brix to 24

Fermentation at 26 C + 2 for 9 days	Addition of wine yeast culture (50%), Separation of wine by cheese cloth
	Filtration
	Scaling of filled bottles
	Pastcurization of 65 C for 20 minutes
	Storage of wine

References:

AMERINE, M.A. BERG, I.I.W. AND CRUNESS, W.V., 1972, The Technology of Wine Making 3rd edition, AVI Publising Company, West Port, Connecticut, PP 126-132

EZYHRINCIW, K, 1966 The Technology of Passion Fruit and Mango Wines, American, Journal of Enology and viticulture, 17:27-33

PROBOGLAV, E.S., ARUTHAYNAN, L.G., MUKHAMEDKANOV, T. AND PEROVA, L.P., 1983, Manufacture of apple-based Acerry Type Wine, International Journal for Food Science and Technology 15:1211-1666.

SHIVANNA, M.R., 1997, Utilization of Jack fruit (Arocarpus Heterophyllus Lam) for Wine Preparation Using Biotropicase Enzyme. M.Sc. (Agri) Thesis, University of Agricultural Sciences, Bangalore.

SINGHNAGI, H .P., AND MANJREKAR S.P., 1975 Studies on the Preparation of Cider from North Indian Apples: II Storage Studies Indian Food Packer, 29:12-15

SURESH, E.R. AND ETHIR. AJ. S. 1988, Studies on the Preparation and Quality Improvement of Wine. Annual Report, Indian Insititute of Horticultural Research Bangalore, P.53.

SURESH, E.R., ETHIRAJ. S. AND NEGL, S.S., 1985 Evaluation of New Grape Cultivars for Prepration of Wine. Journal of Food Science and Technology, 22:211-215

The wine scoring less than 10 marks is considered as poor, between 11 to 15 as better quality and above 15 are treated to be the best quality wines. The dry and sweet wines fetched 15 and 16 points indicate the accuracy of cost effective technology developed for the production of better quality wine. The Vermouth with Cardamam was found to be more acceptable than with clove & ginger.

Wine and health:

1 An occasional wine drinking or a glass, of wine a day is said to stave off near degenerative diseases such as Alzhemers and Parkinsons.

2 Alcolhol concentration in wine may not be anti-microbial but phenolic compounds present in wine are anti-microbial and helps in preventing many diseases like dysentary, diarrhea and have protective effects against cardiovascular diseases.

3 There is positive correlation found between moderate wine consumption and improved health possible due to decrease in the prevalance of age-related mascular degeneration.

4 The consumption of moderate amount of wine can reduce the risk of certain cancers and increased consumption may provoke some cancers.

5 Moderate wine consumers may live on an average 25 years longer than teetotallers and considerably longer than heavy drinkers.

Tips to Wine Drinkers:

1 Most home made wines are good to have, at least a year old. They normally stay well for 3-4 years.

2 Like beer bottle it is not necessary to finish your bottle of wine the day you open it. It keeps well for 2-3 weeks with air tight stoppers.

3 Never keep your wine bottle at room temperature. The ideal temperature is between 12 to 15 degree centigrade.

4 The bottle can simple be kept in the refrigerator

5 Occasional drinking or a glass of wine a day is found to be good for health.

Fermentation:

The pomegranate fruit must is filled in Laboratory fermentor/flast (Plate-3) to its 3/4th capacity and inoculated with 5% wine yeast culture inoculum. The rermentor is plugged with cotton plug to allow sterile air for the initial growth of wine yeast cells. The cotton plug is replaced after 24 hours with air-tight rubber cork fitted with rubber tube. The other end of the rubber tube is kept in water to allow fermentation in the fermeator. The fermentation is better carried out at 21-24°C to get better quality wine.

Siphoning/racking:

After allowing fermentation for 7-8 days, the white coloured and shriveled arils floating in the fermentor is racked with the help of clean muslin cloth to remove pseudo seeds and organic precipitate etc. The fementation is allowed to continue for another two weeks. The clean wine formed is further siphoned into bottles after removing sediments. The bottles can be stored at 15-16°C temperature or simply be kept in refrigerators.

Maturation:

The newly-made wine is harsh in taste and yeasty in flavour. The process of maturation or aging makes the wine tasty and fruity in flavour. After storing in refrigerator for one week racking is done and stored back in the refrigerator. Racking is repeated for 2-3 times at intervals of 15-20 days. Racking can be continued further at monthly intervals. During inter-racking times, no head space should be left in the bottles and should be closed tightly. The matured wine after 6 months to a year can be consumed or pasturized and stored for long time at 15-16°C (Fig-1). The home made wine can stay for 3-4 years.

The detailed production cost of homemade pomegranate wine was worked out and given in Table-1. Vermouth was prepared by addition of different spices like clove, cardamom and ginger at the rate of 0.25, 0.5 and 1.0 g/lit of base sweet wine prepared from fruits of pomegranate. The results of sensory evaluation are presented in Table 3. The results indicated that Vermouth from Pomegranate is found to be acceptable by wine lovers.

The wine prepared from the standard procedure (Plate-4) was subjected to organoleptic evaluation for judging its quality based on sensory properties. From the pomegranate fruits (Arakta cultivar) two types of wines were prepared as dry wine and sweet wine with the inoculation of standard wine yeast of Saccharomyces ellipsoideus No. 101. The scores as per 20 points developed for organolcptic evaluation are given in Table-2 (Amcrine and Cruncss, 1972)

Scope for Indian Wines:

India process over 6.5 million tons of fruits every year and nearly 30-40 per cent fruits may loose their marketability due to undersized fruits, spotted fruits, cracked fruits, immature fruits and post-harvest losses which accounts to be around 2.5 million tonne. Such losses can be avoided by preparing wine out of such fruits. By considering 50 per cent juice from such fruits, it can be converted to wine as value added product for local markets. A good quality wines can be prepared from Pomegranate, Apple, Pineapple and other non conventional fruits.

Methodology of Wine Making:

Preparation of starter culture:

A looful of standard wine yeast culture of Saccharoayces

Ellipsoideus No. 101 maintained on yeast Extract Peptone Dextrose Agar (YEPT) was tranferred into a test-tube containing 5 ml of YEPD broth and incubated overnight at 10°C The 5 ml yeast culture is further added to 100 ml sterilized grape juice (5% inoculum) in 250 ml flask and kept for incubation at 10°C for 22 hours on a shaking incubator at 120 rpm. This can be used to inoculate the fermentor after activating the yeast cells at 10°C for 1.5 hours.

Preparation of Juice/must:

The fruit of pomegranate were crushed, fleshy and juicy arils are separated by hand. The arils are squeezed to get juice, for further wine making.

Amelioration:

The juice (must) should be tested for its sugar content and pH. If the pH of the juice is found less that 3.2 to 3.3 it can be raised by the addition of Sodium bicarbonate. The TSS of pomegranate juice falls in the range of 14-15.* Brix could be raised to 22-30*Brix by the addition of sugar.

Addition of SO_2:

The addition of potassium metabisulphite (KMS) at the rate of 2.00 ppm (100 ppm SO_2) is added to must, so as to eliminate wine yeast. The 5% cultivated wine yeast is added to the must after having 3-4 hours after the addition of KMS. The colour of the juice is temporarily lost, with the addition of KMS it will slowly regain its colour during fermentation period.

(Courtesy Dr. A.B. Patil, Bijapur)

ANNEXURE - I

AEZ - POMEGRANATE

POMEGRANATE EXPORTS FROM INDIA-DURING 1998-99, 1999-2000 & 2000-02

Qty in Kgs. Value in Rs.

Country	1998-99 Quantity	Value	1999-2000	Value	2000-01 Quantity	Value
1	2	3	4	5	6	7
American Samoa	-	-	36	1500		
Australia	-	-	-	-	2000	23493
Belgium	-	-	20800	753480	31200	1027884
Bangladesh	296167	4440822	297506	2529939	140564	224631
Bahrain	194994	35933372	160436	2728963	193174	6421599
Bahamas	-	-	--	18251	475466	
Brazil	20250	1071429	-	-	-	-
Barbodos	-	-	17750	52442	-	-
Canada	36583	10200043	60743	1459593	50040	1284685
Spain	108000	2137085	10950	1408755	-	-
France	35050	783401	400	4351	-	-
Germany	48380	1865189	5698	59823	10000	287100
U.K.	272826	10700621	241554	8829223	161122	6867031
Ghana	600	6530	-	-	-	-
Hong kong	15220	94400	1425	71572	-	
Italy	-	-	-	-	140001	1999200
Japan	5450	196084	10200	448014	11000	377308
Kenya	6600	147833	-	-	-	
Kuwait	125247	2817975	139113	1893082	66625	1162297
LaoPd.Dem. Rep	7740	143871	-	-	-	
Srilanka	237549	4689777	254505	3445519	148362	2464700
Maldives	800	10729	264	8108	-	-
Mexico	20000	1327303	-	-	-	-

Country	1998-99 Quantity	Value	1999-2000	Value	2000-01 Quantity	Value
1	2	3	4	5	6	7
Malaysia	60000	152640	500	24500	-	-
Netherlands	12590	768450	-	-	8000	1534590
Norway	100	2150	-	-	-	-
Nepal	-	-	212	21800	13000	74952
Oman	26890	11177453	49585	795249	182372	4801151
Pakistan	32210	369649	15787	204363	32204	481447
Phillippines	325494	2345108	-	-	-	-
Portugal	35494	2345107	-	-	-	-
Paraguay	1200	24850	-	-	-	-
Qatar	42240	581683	16110a	189261	12387	322862
Rwanda	3110	123719	-	-	-	-
South Africa	-	-	550	26187	47620	805784
Soudi Aribia	790234	11045771	1233470	18613220	170644	3266862
Singapore	12630	320593	22992	482245	1938	37155
Sweden	4709	100430	-	-	-	-
Thailand	1800	85236	-	-	-	-
Trinidad & Tabago	20250	106507	-	-	-	-
Tanzania	300	11497	-	-	-	-
U.A.E.	1758470	35113168	3159530	69523495	292227	62570885
U.S.A.	27860	521328	1750	31983	14575	119513
Product Total	4239148	89606634	5726366	115368664	445541	99155636

Source: APEDA Export Statistics for Agro & Food Products-2000-01 APEDA Regional Office, Hyderabad.

ANNEXURE - II

AEZ-POMEGFRANATE

Production Export of Pomegranate in Important Growing Countries.

Sl. No.	Name of the Country	Variety	Area (in Acres)	Production (in M.T's)	Exports (in M.T's)
1	Afghanistan	N.A.	N.A.	N.A.	N.A.
2	Bangladesh	N.A.	N.A.	N.A.	N.A.
3	Egypt	N.A.	N.A.	N.A.	N.A.
4	China	N.A.	N.A.	N.A.	N.A.
5	Pakistan	N.A.	N.A.	N.A.	N.A.
6	Irna	Malas Maykhosh Alok, taloubarik	1,00,000	6,00,000	20,000
7	Iraq	Ahmar. Aswad. Halwa	N.A.	N.A.	N.A.
8	India	Ganesh, Mridula, Bhagna	1,60,000 Anantput Hyderabad (A.P.) Bijapur, Chitradurga (karnataka) Sangola, Sholapur (Maharashtra)	7,00,000	5,000
9	Isreal Mule Head Wonderful	-	2,500	300	

Sl. No.	Name of the Country	Variety	Area (in Acres)	Produc-tion (in M.T's)	Exports (in M.T's)
10	Saudi Arabia	Mangulati	N.A.	N.A.	N.A.
12	Spain	Mollur deelchi Valcncianas	5,300 (Alicante Province	20,000	10,000
13	USA	Wonderful Graneda	3,500 (California)	17,500	No fresh produce export (except concentrated juice)

Information on other countries will be updated as and when the information is received from the contact sources. Pomegranate is not listed in FAO and other Statistical sources separetely.

LIST OF CHEMICALS REQUIRE COMPULSORY TESTING FOR EXPORT

SlNo.	Chemical	European union	UK	Dutes
Organo Chlorine Chemical				
1	Adrin	Banned	0.05	0.05
2	Chlordane Banned	0.02	0.02	
3	Chloro thalonil	1.00	1.00	1.00
4	DDT	Banned	0.05	0.05
5	Dichloflunid	10.00	15.00	10.00
6	Dicotol	2.00	2.00	2.00
7	Doldrin	Banned	0.05	0.05
8	Endosulfan	0.50	0.50	0.50
9	Endrine	0.01	0.01	0.01
10	Lindane	0.50	0.01	0.01
11	SCH	=	=	0.01
12	Heptachlore	0.01	0.01	0.01
Organophosphorus				
13	Accephate	0.02	0.02	0.02
14	Aggeenphos methyl	1.0	2.0	1.00
15	Chlormotenivphos	0.05	0.05	0.05
16	Chloropyriphos	0.50	0.30	0.50
17	Chlorophyriphosmethyl	0.20	0.20	0.20
18	Dizinon	0.02	0.02	0.02
19	Dichlorovos	0.10	01.10	0.10
20	Dimethoate	1.0	0.02	0.02
21	Fthion	0.50	0.50	0.50
22	Etrimphos	=	=	0.05

SlNo.	Chemical	European union	UK	Dutes
23	Fenhlorophos	0.01	0.01	0.01
24	Fenhlorophos	0.01	0.01	0.01
25	Malathion	0.50	0.50	0.50
26	Profenophos	0.50	0.50	0.50
27	2-Chlorophenol	-	-	-
28	Metharnidphos	0.01	0. 01	0.01
29	Mthedethion	0.50	0.05	0.05
30	Mevinphos	0.10	0.10	0.10
31	Parathionethyl	Bannecd	=	0.05
32	Phosphonidon	0.15	=	0.15
33	Fosmite	Banned	=	0.05
34	Parathion methyl	0.20	0.20	0.20
35	Quinolphos	0.05	0.05	0.05
36	Trizophos	0.02	0.02	0.02
Cynthate parathroid				
37	Gyclothrin	0.30	0.30	0.30
38	Cyclothrin lamda	Banned	=	0.02
39	Cypermethrin	0.50	0.50	0.50
40	Deltamethrin	0.10	0.10	0.10
41	Fenvalarate 0.02	0.02	0.02	
42	Parmithrin	1.00	0.05	1.00
Trizins				
43.	Altrazin asayalannins	0.10	0.10	0.10
44	Banlexil	0.20	0.20	0.20
45	Metalaxil	2.0	2.0	2.0

SlNo.	Chemical	European union	UK	Dutes
Carbamate				
46	Carbofuron	0.10	0.10	0.10
47	Methomyl	0.05	0.05	0.05
Paramidins				
58	Fenarimal	0.30	0.30	0.30
Trizols				
59	Bitertenol	0.05	0.05	0.05
60	Fusiloczol	Banned	Banned	Banned
61	Hexaconozol	0.02	0.10	0.10
62	Propeconizol	0.50	0.50	0.50
Imidozoles				
63	Imizalil	0.02	0.02	0.02
64	Captofol	0.02	0.02	0,.02
65	Captan	3.00	3.00	3.00
Dicorboxy mides				
66	Imrodine	10.00	10.00	10.00
67	prosamidon	5.0	5.0	5.0
68	Vinchlozolin	5.0	5.0	5.0
Benzimizol				
69	Carbandizim	2.0	2.0	2.0
70	Thiophenate methyl	1.0	1.0	1.0
Thiocarbamate Fungicide				
71	Mancozeb	0.01	0.01	0.01
72	Maneb	2.00	2.00	2.00
73	Thiram	0.01	0.01	0.01

SlNo.	Chemical	European union	UK	Dutes
74	propinab	0.01	=	3.00
75	Lthylene Thiourca	0.01	=	0.02
Others				
76	Thiomethaxam Oxydemation methyl	@=0.02	@=0.02	@=0.02
77	Fostylal	#	#	0.20
78	Copper oxychloride	2.00	20.00	20.00
79	Tridemefon	-	2.00	2.00
80	Sulphur	50.00	=	50.00
81	Microbutanil	1.00	1.00	1.00
	Denocap	=	=	0.05
	Cymaczonil	0.01	=	0.05
	Carbaryl	3.00	3.00	3.00

As per Netherland MRI can be used @ synegenta Co. MRL, 0.1 mg kg, but further reduced to 0.02 kmg kg.

(Dept of Horticulture. Maharashtra state) Pune-411005.

Ganesh

Arakta

Ruby

Kesar

Cercospora Leaf Spot

NUTRIENT DEFICIENCIES & REMEDIES
NITROGEN
PHOSPHORUS
POTASSIUM
CALCIUM
MAGNESIUM
SULPHUR
ZINC
IRON
MANGANESE
COPPER
BORON
MOLYBDENUM
multiplex

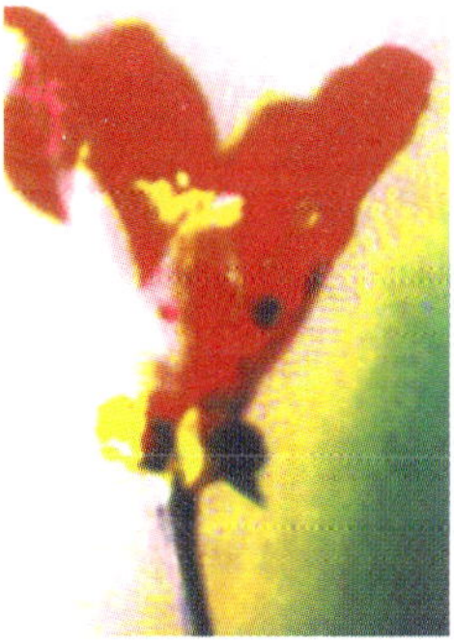

Cercospora Flower Spots

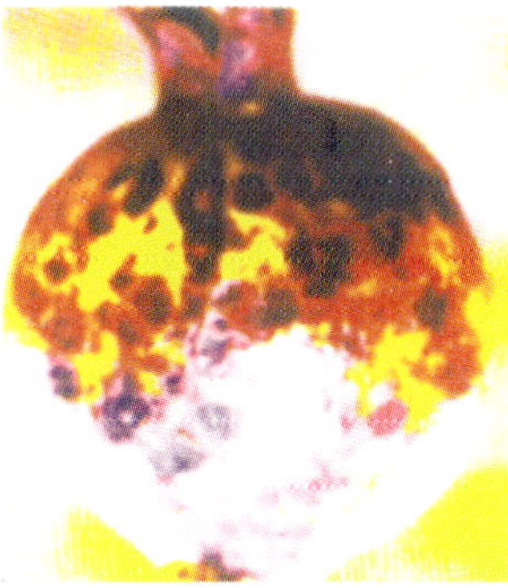

Cercospora Fruit Spots

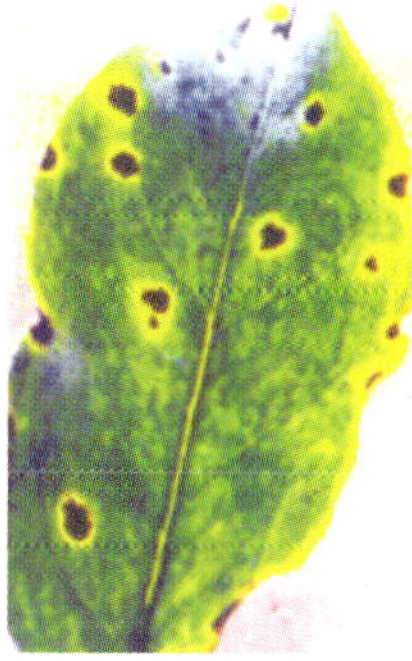

Bacterial Leaf Spots

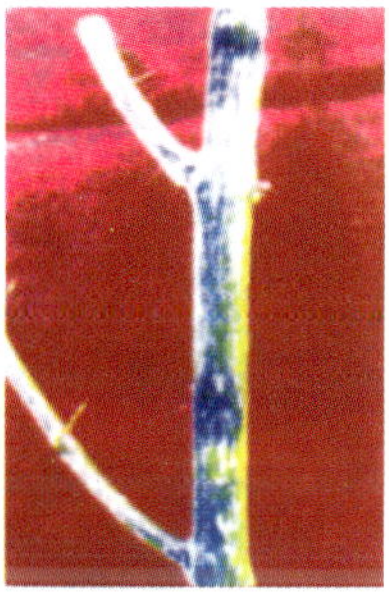

Bacterial Stem Spots

Bacterial Fruit Cracking

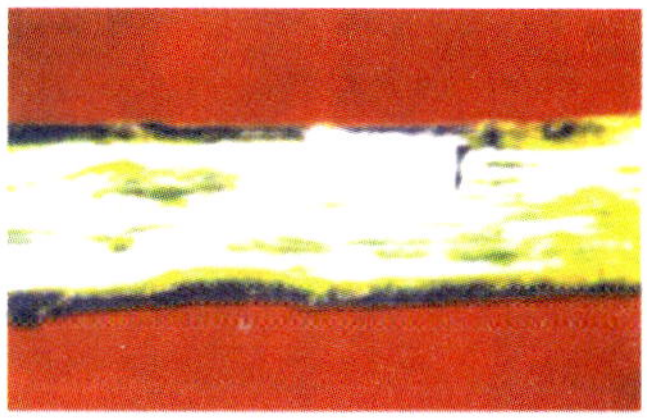

Wilt Spots

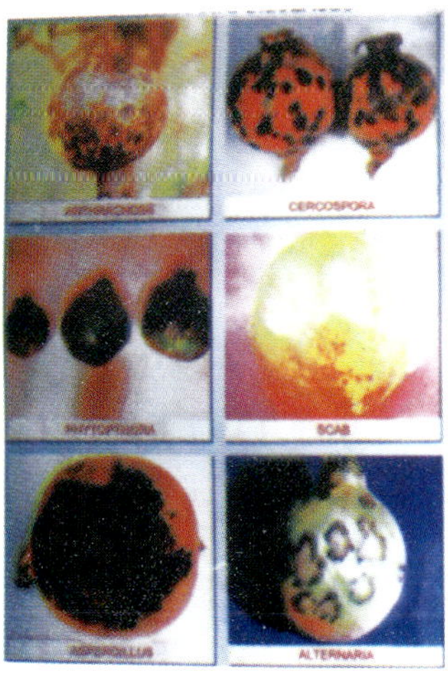

Pomegranate Diseases

Pomegranate Insects

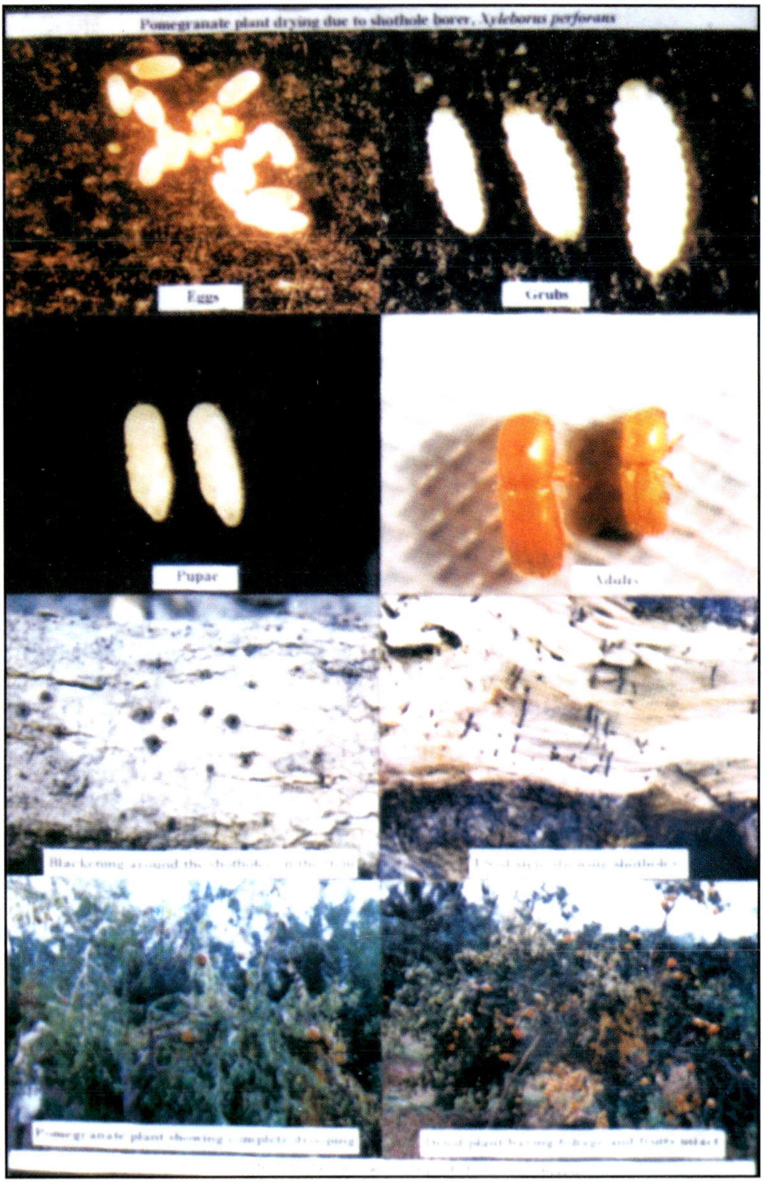

Shot hole borer life cycle and damaging symptoms
(Courtesty S.B. Jagginavar)

Pomegranate Wilt

Scrab Frit Rotting

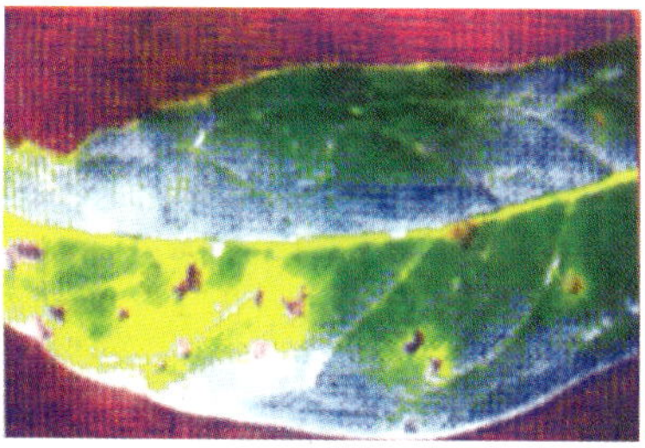

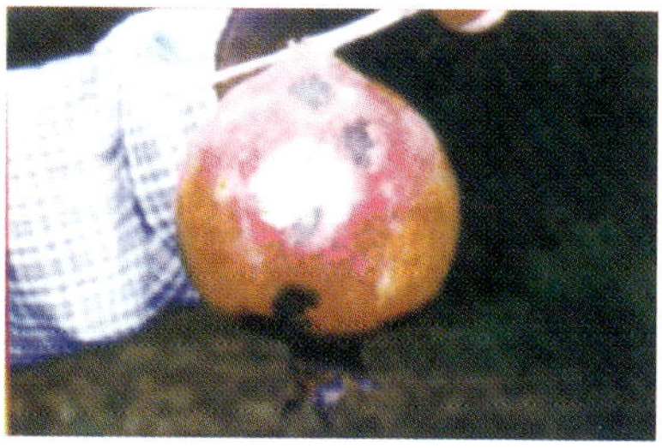

Anthracnose Leaf and Fruit Spot

Thrips Attack

Thrips and Mites

Thrips and Mites Damage

Stem borer damage on twigs

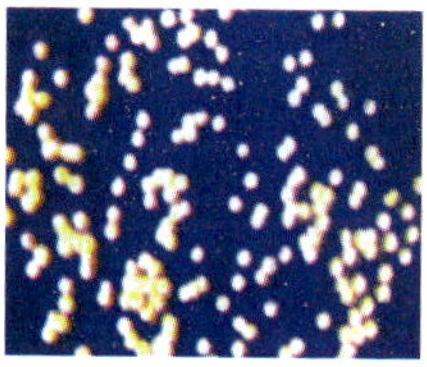

Moth Eggs

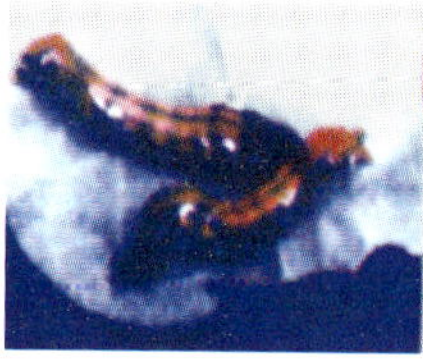

Moth Larva

Moth Adult

Moth Sucking Juice

Moth Damage

Mealy Bug

Pomegranate under net

Fruit Borer

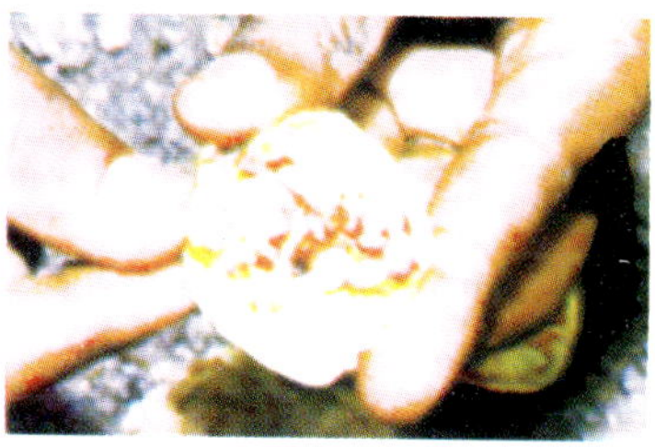

Fruit Borer Rotting

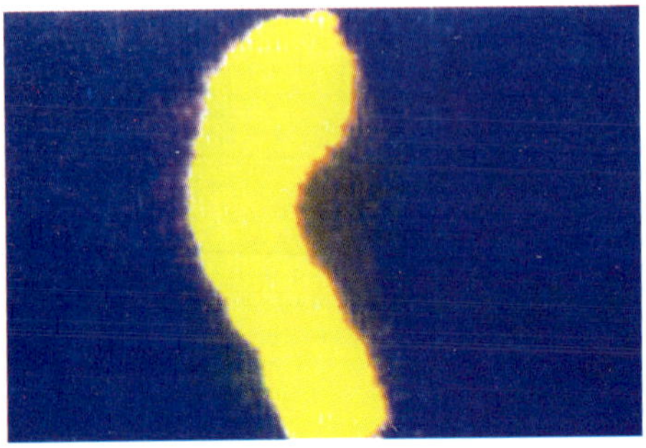

Stem Borer Grub

Stem Borer damage with excreta below the plant

Stem Borer Damage

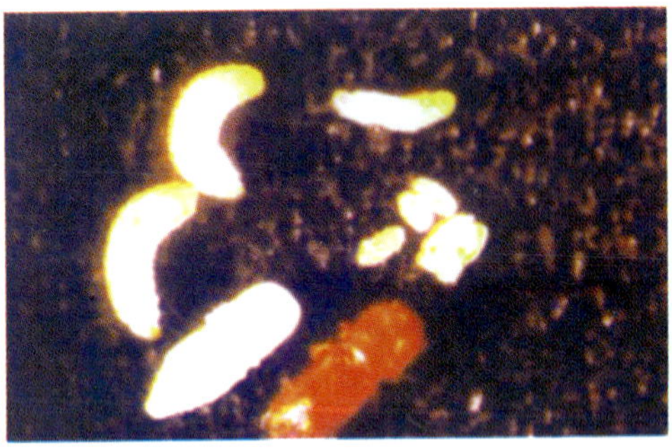

Shot Hole borer Egg, Grub, Adult Stages

Stem borer Adult

Shot Hole borer, Damage symptoms

Index